Self-Assembled Bio-Nanomaterials

Self-Assembled Bio-Nanomaterials

Synthesis, Characterization, and Applications

Special Issue Editor

Gang Wei

MDPI • Basel • Beijing • Wuhan • Barcelona • Belgrade

Special Issue Editor
Gang Wei
Qingdao University
China

Editorial Office
MDPI
St. Alban-Anlage 66
4052 Basel, Switzerland

This is a reprint of articles from the Special Issue published online in the open access journal *Nanomaterials* (ISSN 2079-4991) from 2018 to 2019 (available at: https://www.mdpi.com/journal/nanomaterials/special_issues/self_assem_bionanometer_biosynth_charact_appl).

For citation purposes, cite each article independently as indicated on the article page online and as indicated below:

LastName, A.A.; LastName, B.B.; LastName, C.C. Article Title. *Journal Name* **Year**, *Article Number*, Page Range.

ISBN 978-3-03928-536-5 (Pbk)
ISBN 978-3-03928-537-2 (PDF)

Contents

About the Special Issue Editor

Gang Wei received his Ph.D. from the Changchun Institute of Applied Chemistry, Chinese Academy of Sciences, China, in 2007. Since 2012, he has worked as a senior researcher and group leader in Hybrid Materials Interfaces Group at the University of Bremen, Germany. In October 2019, he was appointed as a full professor at the Qingdao University, China. He has obtained the Fellows of the Alexander-von-Humboldt (2007), Carl-Zeiss (2009), and Young Taishan Scholars (2019) Programs. His research interests include two-dimensional nanomaterials, supramolecular self-assembly, biomaterials, sensors, and biosensors, as well as single-molecule force spectroscopy. He has published 130+ papers in peer-reviewed journals such as *Chem Soc Rev*, *Prog Polym Sci*, *Adv Funct Mater*, *ACS Nano*, and others, and the published papers have been cited more than 4700 times with an H-index of 43.

Preface to "Self-Assembled Bio-Nanomaterials"

Supramolecular self-assembly is a simple but effective bottom-up technique for creating functional nanomaterials with novel structures and properties. Biomolecules, such as DNA, proteins, peptides, virus, enzymes, biopolymers, and others, have unique abilities to form hierarchical and ordered 1D, 2D, and 3D nanostructures and nanomaterials by molecular self-assembly in liquid, solid surfaces, and air–water interfaces. The self-assembled bio-nanomaterials have shown wide applications in the fields of biomedical engineering, tissue engineering, biosensors, nanotechnology, energy materials, and others. For example, self-assembled peptide and protein bio-nanomaterials have been used to deliver drugs into the body system with specific targeting, and biomineralized self-assembled nanomaterials have shown excellent potential to repair old tissues and as substitutes for natural structures and functions. By combining self-assembled bio-nanomaterials with other functional nanomaterials like nanoparticles, carbon nanotubes, and grapheme together, hybrid bio-nanomaterials with multiple functions can be created. Compared with inorganic nanomaterials, the formed hybrid bio-nanomaterials have high biocompatibility, self-assembly ability, physicochemical stability, and molecular recognition ability, providing various potential applications in cell labelling, bioimaging, biosensors, and functional materials. Many achievements have been occurred in this interesting field; however, it is still necessary and significant to investigate: (i) the synthesis of novel bio-nanomaterials by molecular self-assembly of various biomacromolecules and small molecules, (ii) the multi-characterizations of created biomaterials, and (iii) the potential applications of both pure self-assembled and hybrid bio-nanomaterials in drug delivery, regenerative medicine, tissue engineering, cell culture, bioimaging, sensors, functional materials, and others.

Gang Wei
Special Issue Editor

nanomaterials

Review

Controlling the Self-Assembly of Biomolecules into Functional Nanomaterials through Internal Interactions and External Stimulations: A Review

Li Wang [1,*], Coucong Gong [2], Xinzhu Yuan [1] and Gang Wei [2,*]

1 Key Laboratory of Preparation and Application of Environmental Friendly Materials (Jilin Normal University), Ministry of Education, Changchun 130103, China; 15144435516@163.com
2 Faculty of Production Engineering, University of Bremen, D-28359 Bremen, Germany; ccgong@uni-bremen.de
* Correspondence: liwang@jlnu.edu.cn (L.W.); wei@uni-bremen.de (G.W.); Tel.: +49-421-218-64581 (G.W.)

Received: 28 January 2019; Accepted: 15 February 2019; Published: 18 February 2019

Abstract: Biomolecular self-assembly provides a facile way to synthesize functional nanomaterials. Due to the unique structure and functions of biomolecules, the created biological nanomaterials via biomolecular self-assembly have a wide range of applications, from materials science to biomedical engineering, tissue engineering, nanotechnology, and analytical science. In this review, we present recent advances in the synthesis of biological nanomaterials by controlling the biomolecular self-assembly from adjusting internal interactions and external stimulations. The self-assembly mechanisms of biomolecules (DNA, protein, peptide, virus, enzyme, metabolites, lipid, cholesterol, and others) related to various internal interactions, including hydrogen bonds, electrostatic interactions, hydrophobic interactions, π–π stacking, DNA base pairing, and ligand–receptor binding, are discussed by analyzing some recent studies. In addition, some strategies for promoting biomolecular self-assembly via external stimulations, such as adjusting the solution conditions (pH, temperature, ionic strength), adding organics, nanoparticles, or enzymes, and applying external light stimulation to the self-assembly systems, are demonstrated. We hope that this overview will be helpful for readers to understand the self-assembly mechanisms and strategies of biomolecules and to design and develop new biological nanostructures or nanomaterials for desired applications.

Keywords: self-assembly; biomolecules; nanostructures; interactions; external stimulations

1. Introduction

Self-assembly is a simple but effective bottom-up technique for preparing functional nanomaterials with ordered structures and novel functions [1–3]. Besides nanoparticles (NPs) [4], polymers [5], and other inorganic nanoscale building blocks, many kind of biomolecules in nature, including DNA [6], proteins [7], peptides [8,9], viruses [10,11], enzymes [12,13], and others [14], have also exhibited great potential to form hierarchical nanomaterials by controllable self-assembly. Due to the unique molecular properties, adjustable functions, and ordered structures, the self-assembled biological nanomaterials have been widely utilized for applications in the fields of materials science, biomedical engineering, tissue engineering, biosensors, and nanotechnology [15–20].

In order to fabricate functional biological nanomaterials, one of the key challenges is how to control the self-assembly of biomolecules to form desired structures. Previous studies have indicated that this challenge could be solved through adjusting internal molecule–molecule/materials interactions (such as hydrogen bonding, electrostatic interaction, hydrophilic/hydrophobic interaction, and DNA/RNA hybridization [21–24]) or carrying out external stimulations (such as adjusting the pH, temperature, or ionic strength or adding organics and enzymes to the system [25–28]). To further improve the functions and applications of self-assembled biomolecular nanomaterials, some functional nanoscale building blocks, such as nanoparticles [29,30], carbon nanotubes [31], graphene [32,33], and polymers [34,35], could be introduced into the biomolecular self-assembly systems, where the potential biomolecule–materials (building blocks) interactions could guide the self-assembly of both biomolecules and the corresponding building blocks into hybrid nanomaterials. For instance, Yu et al. demonstrated the biomolecule-assisted self-assembly of CdS/MoS_2/graphene hollow spheres for high-efficiency and low-cost photocatalysis [30]; Zou and co-workers reported the fabrication of novel nanodots by the self-assembly of peptide–porphyrin conjugates [35].

The self-assembly mechanisms of biomolecules to various nanostructures have been investigated widely, and some reviews on the design, synthesis, and applications of self-assembled biomolecular nanomaterials have been reported previously [36–40]. For example, Yang and co-workers provided an overview on the self-assembly of proteins to various supramolecular materials, in which the design strategies for self-assembling proteins were introduced and discussed in detail [39]. Willner et al. summarized the applications of biomolecule-based nanostructures and nanomaterials for sensing and the fabrication of nanodevices [40]. After studying these reports, we realized that it is still valuable for us to contribute a review on the self-assembly of biomolecules to functional nanomaterials from the viewpoints of internal interaction mechanisms, external stimulation, and designed functionalities.

In this review, we focus on the fabrication of biological nanomaterials by controlling the self-assembly of a biomolecule through internal biomolecular interactions and external stimulations. The main used/studied biomolecules, including proteins [7,41], peptides [42,43], amphiphiles [44], DNA [45], carbohydrates, metabolites [46–48], lipids, and cholesterol, for the self-assembly of various nanostructures are introduced and discussed in detail. In Section 2, some recent studies are analyzed and discussed to illustrate the self-assembly mechanisms of various biomolecules, in which the strategies for creating biological nanomaterials via internal interactions are presented. In Section 3, we introduce recent studies on adjusting biomolecular self-assembly via external stimulations. In Section 4, we provide a summary on the synthesis of biological nanomaterials based on various biomolecules. It is expected that this work will be helpful for readers to understand the self-assembly mechanisms of, and strategies for creating, functional nanomaterials, and to design and develop new biological nanostructures and nanomaterials for advanced applications in materials science, biomedical engineering, analytical science, energy, and environmental science.

2. Internal Interactions towards Biomolecular Self-Assembly

The mechanisms of biomolecules self-assembled into various nanostructures are complex. In this section, we demonstrate the self-assembly mechanisms of biomolecules from basic molecular interactions (such as hydrogen bonds and electrostatic, hydrophobic, and π–π interactions) as well as biomolecular-specific interactions (such as DNA base pairing, ligand–receptor/antigen–antibody binding, and biomolecule–polymer conjugates).

2.1. Basic Molecular Interactions

2.1.1. Hydrogen Bonds

Hydrogen bonds, the interactions that occur between hydrogen atoms and a great number of electronegative atoms in biomolecules, play important roles in the formation of biological nanomaterials [49–51]. In the process of biomolecular self-assembly, the hydrogen bonds can promote the growth of biomolecules in one direction with a long-range order to form one-dimensional (1D) nanostructures. In addition, the hydrogen bonds between the hydrogen atoms of biomolecules and the electronegative atoms of a special material's surface could enhance the interactions between biomolecules and materials for the formation of functional biomolecule-based hybrid nanomaterials.

Diphenylalanine (FF) is a popular dipeptide motif for self-assembly in water driven by hydrophobic interactions; however, other interactions also likely play a role. For instance, Li et al.

demonstrated the formation of FF microrods by hydrogen-bond-based self-assembly [21]. The structural and property characterizations of the self-assembled microrods indicated that 1,1,3,3,6,6-Hexafluoro-2-propanol (HFP) formed stable intermolecular hydrogen bonds with an FF peptide, leading to the solvation of peptide molecules. When the peptide was dropped onto a silicon wafer, the evaporation of HFP promoted the self-assembly of FF to form nanofibers, microtubes, and microrods, as shown in Figure 1a. It can be found that FF molecules first self-assemble into nanotubes in the presence of water, and then grow into microrods through both hydrogen bonds and hydrophobic interactions between aromatic residues of the peptide. In another study, Yang and co-workers investigated the self-assembly of an FF peptide on a graphene surface and the formation of peptide nanowires (PNWs) [52]. Firstly, the peptide solution was diluted with a certain concentration of graphene dispersion solution and then dropped onto a clean substrate to dry in an oven at 50 °C, as shown in Figure 1b. The self-assembly of the FF peptide with graphene in water was ascribed to both hydrogen bonds and π–π interactions. It can be concluded that graphene promoted the π–π conjugations between peptides and graphene, and the intermolecular/molecule–graphene hydrogen bonds mediated the formation of ordered PNW arrays on the graphene's surface.

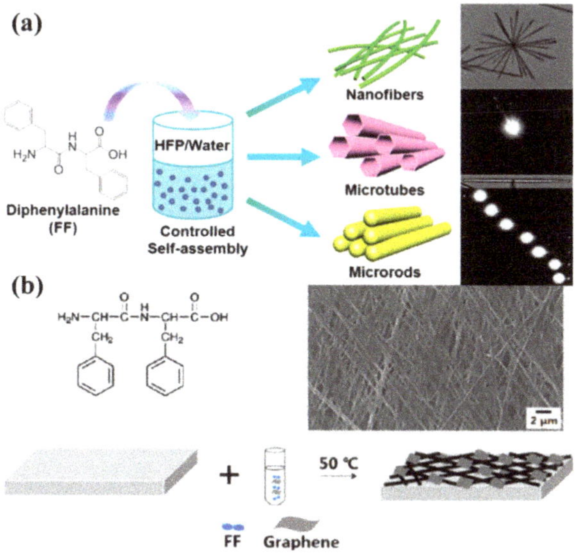

Figure 1. Hydrogen bonds promoted the self-assembly of biomolecules: (**a**) Hydrogen-bond-induced self-assembly of FF into nanofibers, microtubes, and microrods. Reprinted with permission from [21]. Copyright 2015 American Chemical Society. (**b**) The formation of microscale peptide nanowires (PNWs)–graphene array. Reprinted with permission from [52]. Copyright 2013 American Chemical Society.

Besides peptides, proteins, enzymes, DNA, and viruses can also be utilized to form self-assembled biological nanomaterials through hydrogen bonds. For example, Lee et al. fabricated a ultrathin nanomesh membrane based on the self-assembly of M13 virus on graphene oxide (GO) nanosheets via both hydrogen bonds and electrostatic interactions [53]. In the self-assembly process of virus on the GO's surface, the basic amino acids (histidine and lysine) of the virus enable strong electrostatic and hydrogen bond interactions with the carboxylate groups at the edges of GO nanosheets. In another case, Xue et al. investigated the binding of DNA with GO through the surface plasmon resonance (SPR) technique [54], and found that the hydrogen bond plays a key role in the interactions between single-stranded DNA (ssDNA) with GO, which enabled the fabrication of a novel biosensor for highly sensitive and selective determination of ssDNA targets.

2.1.2. Electrostatic Interaction

Electrostatic interactions play significant roles in the self-assembly of peptides, proteins, enzymes, and others into higher hierarchical nanostructures and at the same time stabilize the formed nanostructures [55–57].

Wang et al. investigated the self-assembly of a motif-designed peptide for the formation of peptide nanofibers (PNFs) and bioinspired PNF-based silver nanowires (AgNWs), and fabricated graphene nanosheet (GN)-PNF-AgNW nanocomposites through an electrostatic interaction between negatively charged PNF-AgNWs and a polymer-modified, positively charged GN [58]. Firstly, a positively charged GN was obtained with a zeta potential of +35.87 mV when a GN was functionalized with the positively charged polymer PDDA. Subsequently, the PNF-AgNW nanohybrids were created by controlling the self-assembly of a designed peptide molecule and subsequent bioinpired synthesis of AgNPs on PNFs, and the formed PNF-AgNW nanohybrids exhibited a zeta potential of −16.39 mV. Therefore, the PNF-AgNW nanohybrids were bound onto PDDA-GN by the electrostatic interaction easily, as indicated in Figure 2a.

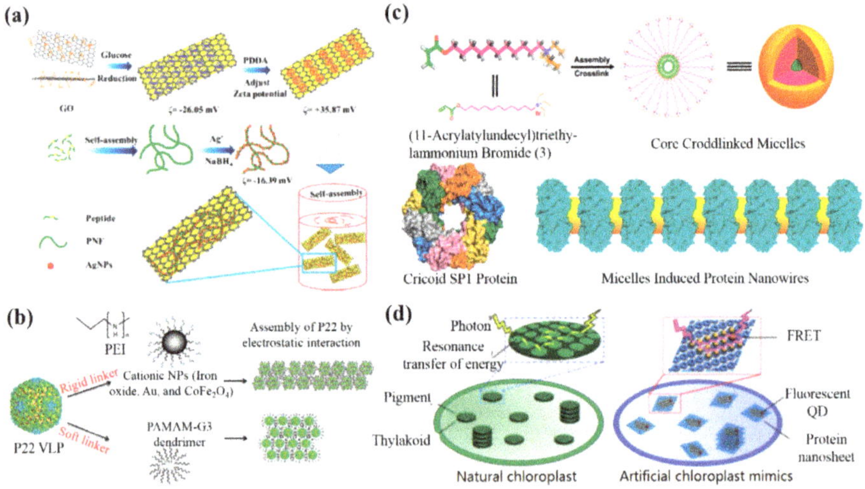

Figure 2. Electrostatic-interaction-mediated self-assembly of biomolecule-based nanomaterials: (**a**) The formation of peptide nanofiber (PNF)-bioinspired silver nanowires (AgNWs) on graphene nanosheets (GNs). Reprinted with permission from [58]. Copyright 2014 American Chemical Society. (**b**) The synthesis of self-assembled P22 virus-like particles (VLPs) via rigid (inorganic nanoparticles (NPs)) and soft (PAMAM) cationic linkers. Reprinted with permission from [59]. Copyright 2017 Materials Research Society. (**c**) The structure-based design of protein nanowires. Reprinted with permission from [60]. Copyright 2016 American Chemical Society. (**d**) Protein nanosheet–quantum dot (QD) nanohybrids. Reprinted with permission from [61]. Copyright 2017 American Chemical Society. PEI, polyethyleneimine; FRET, fluorescence resonance energy transfer.

Gupta and co-workers demonstrated the inorganic NP ($CoFe_2O_4$ and Au)-induced self-assembly of bacteriophage P22 via an electrostatic interaction [59]. In their work, negatively charged P22 virus-like particles (VLPs) were synthesized by the co-expression of the coat protein and scaffold protein in *Escherichia coli*. The positively charaged $CoFe_2O_4$ and Au NPs were prepared by coating the NPs with polyethyleneimine via a modified bilayer phase transfer method. Finally, the P22 VLPs were assembled through a controllable electrostatic interaction between the negatively charged VLP and the positively charged polymer-modified NPs, as shown in Figure 2b. In another study, Liu and co-workers fabricated micelle-induced protein nanowires via an electrostatic interaction when the electronegative cricoid stable protein one (SP1) assembled with positively charged core-crosslinked micelles (Figure 2c) [60]. Recently, they further used SP1 as a building block to assemble positively charaged semiconductor quantum dots (QDs) via an electrostatic interaction to obtain highly ordered protein nanowires with prominent optical properties (Figure 2d) [61].

2.1.3. Hydrophobic Interaction

Many biomolecules, such as peptides and proteins, can form highly ordered self-assembled superstructures via a hydrophobic interaction due to their hydrophobic property [62]. It is well-known that amino acids can be divided into hydrophobic and hydrophilic ones due to their amino acid residues. To date, a lot of studies on the hydrophobic-interaction-induced self-assembly of proteins and peptides for functional bionanomaterials have been reported [63,64].

For instance, Liao et al. investigated the self-assembly mechanism of PNFs in solution and on a surface by using a small peptide amphiphile (PA) (NapFFKYp) as a model [65]. It was found that this PA first grows into nanofibers via a nucleation process, and then forms highly ordered nanofibers in solution by a hydrophobic interaction. However, the self-assembly of this PA could form mixed nanofiber and nanosheet structures on a substrate. Further molecular dynamics simulations (MDSs) suggested that both hydrophobic and ion–ion interactions are crucial during the self-assembly process of this PA. In another study, Yang and co-workers demonstrated how a model ionic-complementary peptide EAK16-II (AEAEAKAKAEAEAKAK) assembles on hydrophilic (mica) and hydrophobic (HOPG) substrates via electrostatic and hydrophobic interactions, respectively [66].

Biomolecules can also be conjugated with other nanomaterials, such as graphene or NPs, to form functional nanomaterials via a hydrophobic interaction [67–70]. For example, Lu and co-workers developed a novel fluorescent approach to monitor peptide–protein interactions based on the assembly of a pyrene-labeled peptide on GO via both π–π and hydrophobic interactions [71]. To achieve the aim, the peptide was firstly modified with a pyrene group to form a π-rich framework with high fluorescence, and then the pyrene-labelled peptide was mixed with GO to obtain a GO-pyrene-peptide nanocomposite through both π–π and hydrophobic interactions, as shown in Figure 3a. Due to the peptide–protein interaction, the competitive binding of an antibody with GO for the pyrene-labelled peptide decreased the adsorption of the peptide on the GO and promoted the formation of a peptide–antibody complex in solution (Figure 3a).

In another study, Ma et al. demonstrated a facile strategy to prepare protein-based NPs [72], where bovine serum albumin (BSA) was modified with multi-photoinitiated reversible addition-fragmentation chain transfer (RAFT) polymerization to the BSA–PHPMA conjugates. The synthesized BSA–PHPMA conjugates were further aggregated into NPs through the hydrophobic interaction of PHPMA (Figure 3b). Zhang et al. reported the immobilization and self-assembly of horseradish peroxidase (HRP) and oxalate oxidase (OxOx) on chemically reduced graphene oxide (CRGO) [73]. Their results indicated that the enzymatic loading can be improved by increasing the reduction degree of GO, and the excellent properties of the CRGO–enzyme conjugates are attributed to hydrophobic interactions between enzymes and the CRGO's surface. Studies on the self-assembly of proteins and enzymes on a material's surface are helpful for designing and fabricating novel biosensors for the high-performance sensing of various analytes.

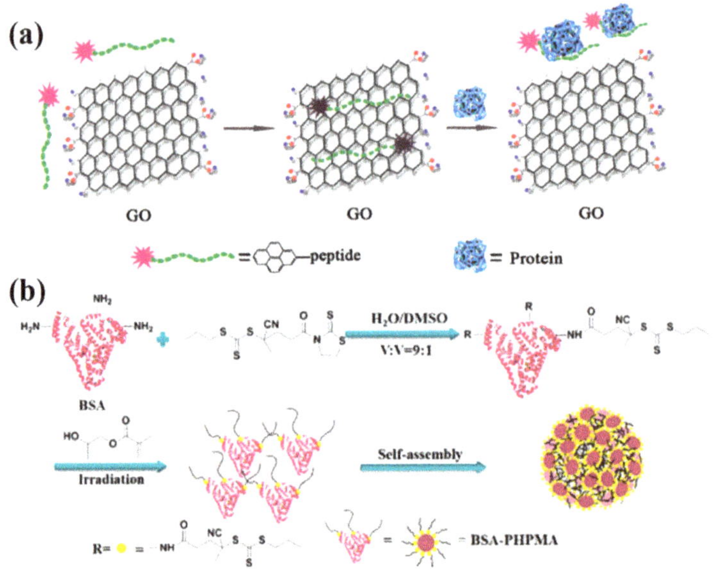

Figure 3. The hydrophobic interaction for biomolecular self-assembly: (**a**) A pyrene-labeled peptide for monitoring the protein–peptide interactions. Reprinted with permission from [71]. Copyright 2011 American Chemical Society. (**b**) Self-assembly of bovine serum albumin (BSA)-based nanoparticles to microspheres. Reprinted with permission from [72]. Copyright 2017 American Chemical Society. GO, graphene oxide.

2.1.4. π–π Interaction

Noncovalent π–π interactions are another potential driving force to promote the self-assembly of biomolecules [74,75]. Some biomolecules, including peptides, proteins, DNA, enzymes, and viruses, mostly contain aromatic motifs, making it possible to form highly ordered superstructures by π–π stacking or functional hybrids by biomolecule–material π–π interaction.

For example, Su and co-workers investigated the self-assembly of a designed peptide (RGDAEAKAEAKYWYAFAEAKAEAKRGD) to PNFs and the π–π conjugation between PNFs and graphene quantum dots (GQDs) towards novel PNF–GQD nanohybrids for the simultaneous targeting and imaging of tumor cells [76]. The designed peptide has trifunctional motifs, in which RGD can recognize the integrin-rich tumor cells, the AEAKAEAK motif provides the capability of self-assembly and formation of PNFs, and the motif of YWYAF has the tendency to bind with GQDs via π–π interactions (Figure 4a). In another study, they synthesized GO–PNF nanohybrids via π–π interactions between designed PNFs and GO, and further utilized the formed GO–PNF nanohybrids as templates for the biomimetic mineralization of hydroxyapatite (HA) (Figure 4b) [33]. Recently, they also used the peptide (AEAKAEAKYWYAFAEAKAEAK) to synthesize GQD–PNF–GO nanohybrids via π–π interactions between the created PNFs with GQDs and GO [77], as shown in Figure 4c. To prove the unique interactions among the three components, they used the atomic force microscopy (AFM)-based force spectroscopy technique to measure the interactions (including π–π binding forces) between PNFs with GQDs and GO, and the obtained results showed that the rupture force between PNFs and GO was stronger than the force between GQDs and GO as well as between GQDs and PNFs.

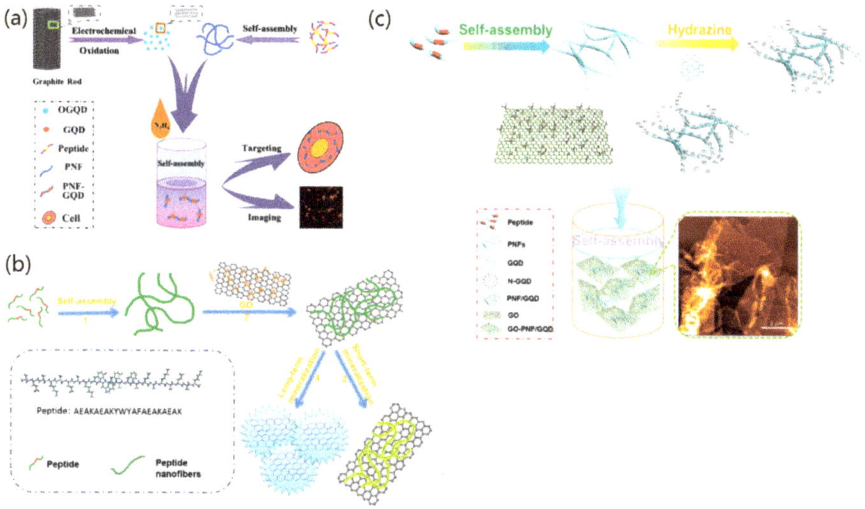

Figure 4. π–π-interaction-mediated self-assembly of nanomaterials: (**a**) The fabrication of PNF–graphene quantum dot (GQD) nanohybrids. Reprinted with permission from [76]. Copyright 2015 WILEY-VCH. (**b**) The synthesis of GO–PNF nanohybrids and GO-PNF-HA minerals. Reprinted with permission from [33]. Copyright 2015 Elsevier. (**c**) The synthesis of PNFs and binary GQD-PNF, and ternary GQD–PNF–GO nanohybrids. Reprinted with permission from [77]. Copyright 2017 WILEY-VCH.

Besides peptides, other biomolecules, such as proteins, DNA, enzymes, and viruses, have also been widely used to conjugate with graphene to form nanomaterials via π–π interactions for various applications [78,79]. For example, Wang and co-workers designed a uniform three-dimensional (3D) graphene-nanodots-encaged porous gold electrode for loading enzyme [80]. Pyrene-functionalized glucose oxidase (GOx) and catalase (CAT) were prepared and used as the immersed solution for the modification of the electrode. In the process of enzyme modification, Pyrene-GOx/CAT was loaded onto the graphene-nanodots-encaged porous gold electrode via π stacking between pyrene and graphene. The fabricated enzyme electrodes showed an excellent catalytic performance compared with native enzyme. Huang et al. designed a GQDs–ionic liquid–nafion (GQDs-IL-NF) composite film [81], which could interact with ssDNA through noncovalent π–π interactions to fabricate a novel biosensor platform for detecting a carcinoembryonic antigen with high sensitivity.

By using the π–π interaction between ssDNA and RGO, Li et al. created ssDNA–RGO composites for the further bioinspired synthesis of cotton-flower-like platinum nanoparticles (PtNPs) [82]. The ssDNA molecules were conjugated with RGO via the π–π interaction and PtNPs were formed on the surface of ssDNA–RGO by the bioinspired synthesis. The created ssDNA-RGO-PtNPs material exhibited high catalytic activity for methanol oxidation and CO tolerance.

2.2. *Biomolecular-Specific Interactions*

2.2.1. DNA/RNA Base Pairing

The focus of DNA/RNA nanotechnology is to direct ssDNAs/ssRNAs to self-assemble into desired 1D, two-dimensional (2D), and even 3D nanomaterials via DNA/RNA base pairing.

Previously, it has been reported that DNA nanostructures could be created by controlling the DNA base pairing through several methods, including the clamped hybridization chain reaction [24], paranemic crossover DNA motif assembly [83], DNAzyme-based logic gate [84], and genetic encoding [85]. The basic principles for creating functional DNA nanomaterials are based on the design of DNA motifs/tiles and the subsequent controllable DNA base pairing. In a typical study, Elbaz and co-workers investigated the fabrication of DNA nanostructures by guiding the self-assembly of gene-encoded DNA in living bacteria [85]. To achieve this aim, a short ssDNA was first encoded as a gene and then enzymatically converted to a new ssDNA. This in-vivo-produced ssDNA could be utilized for in-vitro fabrication of 1D nanowires (Figure 5a,b) and 2D nanosheets (Figure 5c and d). To form a Z- and C-shaped tile, two red pairs of a symmetric motif hybridized with two other red pairs of another symmetric motif. The motif with a short black domain mediated the formation of the Z-shaped (Figure 5a) tile and the motif with a long black domain promoted the formation of the C-shaped tile (Figure 5c). This work provided a novel strategy to prepare functional DNA nanostructures for in-vivo applications.

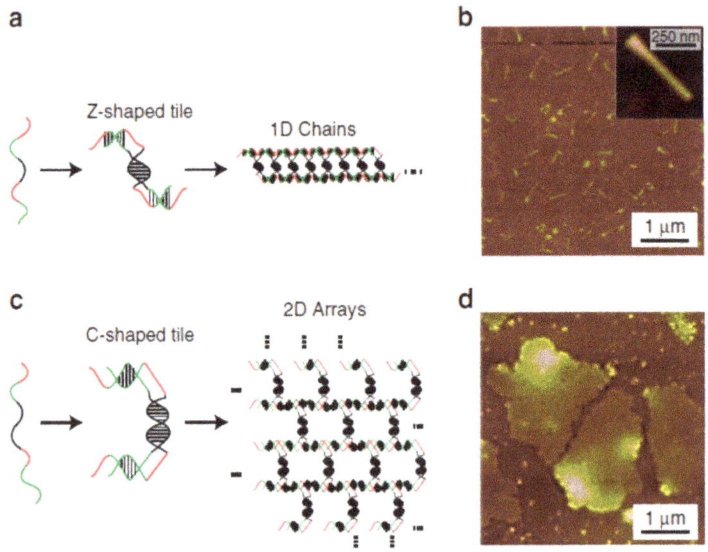

Figure 5. The in vitro genetic-encoding-mediated self-assembly of DNA to different nanostructures: (**a**,**b**) nanowires and (**c**,**d**) two-dimensional (2D) nanosheets. Reprinted with permission from [85]. Copyright 2016 Macmillan Publishers Limited. 1D, one dimensional.

Besides the simple DNA nanostructures, DNA nanotechnology provides the possibility for the programmable design and synthesis of complex nanostructures, such as DNA origami [86,87] and nanoswitches [88]. For instance, in the DNA origami technique, long ssDNA is folded into target shapes by using short DNA staples, which are designed to be complementary to particular regions of the long DNA [86]. Therefore, a lot of 2D and 3D DNA nanostructures can be created by using the further self-assembly of DNA origami and nanoswitches [86,87], which can serve as nanoscale templates to form biomolecule and NP-based nanomaterials [89,90].

Similar to DNA base pairing, RNA can also be assembled into various nanostructures (such as tetrahedrons, nanotriangles, lattices, and tubes) by bottom-up self-assembly based on intra- and inter RNA interactions [91–93]. These self-assembled RNA nanostructures have shown wide applications for drug or NPs delivery and cancer diagnostics [91,93].

Previously, Gazit and co-workers have prepared self-assembled peptide nucleic acid (PNA) fibers with a unique light-emitting property [94], in which the self-assembly of PNA is related to both π–π stacking interactions and Watson–Crick base pairing.

2.2.2. Ligand–Receptor Binding

It is possible to utilize specific molecule–molecule recognitions, such as the ligand–receptor [95–97] and antigen–antibody [98,99] bindings, to guide the self-assembly of biomolecules to form ordered nanostructures and nanomaterials.

Liljeström et al. investigated the self-assembly and modular functionalization of 3D cowpea chlorotic mottle virus (CCMV) crystals by using the avidin–biotin recognition [95]. The functionalization and self-assembly of 3D CCMV crystals were achieved in two ways, as shown in Figure 6. In the first way (Method 1), the functionalization was achieved by first modifying avidin with biotin-linked functional units (such as dyes, enzymes, or NPs), and then adding CCMV particles to form 3D crystals through self-assembly. In another method (Method 2), the mixing of CCMV and avidin together caused the formation of 3D crystals via self-assembly, which could be further functionalized with biotin-linked functional units to form functional 3D nanomaterials. The use of avidin–biotin binding in this study allowed for the highly selective functionalization of protein crystals, which exhibited great potential for biomedical applications. In another study, Xu and co-workers demonstrated that a ligand–receptor interaction could modulate the energy landscape of peptide self-assembly and affect the formation of various nanostructures [96]. Their study proved that it is possible to use a ligand–receptor interaction to modulate the kinetics of enzyme-mediated peptide self-assembly.

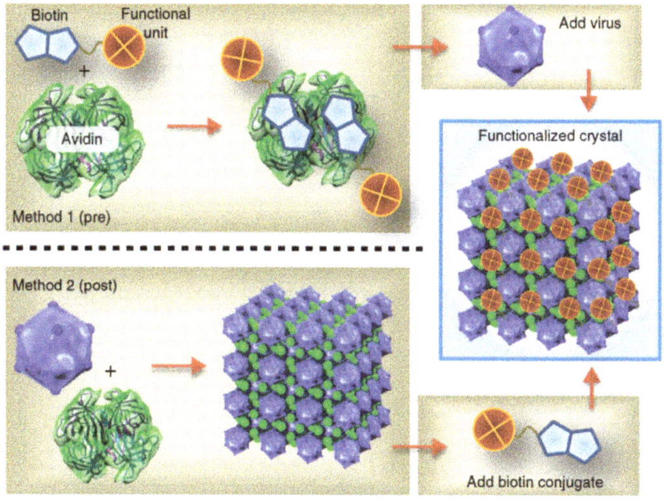

Figure 6. The avidin–biotin-binding-mediated self-assembly of a protein cage to three-dimensional (3D) functional crystals. Reprinted with permission from [95]. Copyright 2014 Macmillan Publishers Limited.

The formation of antigen–antibody immunocomplexes is helpful for the molecular self-assembly and the formation of functional bionanomaterials. For instance, Kominami and co-workers investigated the self-assembly of immunoglobulin G (IgG) on a mica surface, and found the formation of 2D hexameric IgG crystal [99]. Therefore, the antigen (anti-human serum albumin) could be bound onto the IgG hexamer via the antigen–antibody binding.

2.2.3. Biomolecule–Polymer Conjugates for Self-Assembly

The conjugation between biomolecules (such as a lipid and cholesterol) with polymers can also mediate the self-assembly of the formed conjugates to various nanostructures [100,101]. For instance, the mixing of lipid particles with amphiphilic, hydrophobic, and hydrophilic drugs for the formation of hybrid cubosomes and hexosomes has exhibited great potential for advanced drug delivery systems [100,102,103].

Cholesterol, one of the important biopolymers, has been utilized as a versatile building block to conjugate with various polymers for the fabrication of self-assembled functional nanomaterials [104–106]. For example, Yang and co-workers demonstrated the preparation of polyoxometalate–cholesterol conjugates, which could self-assemble into microrods and nanoribbons by controlling the reaction temperature [104]. Engberg et al. found that cholesterol could be tethered into poly(ethylene glycol) (PEG) networks, via polymerization in an organic solvent, that were capable of forming weakly ordered aggregates via self-assembly [105].

3. External Stimulations towards Biomolecular Self-Assembly

The self-assembly of biomolecules is highly sensitive to, among other things, the molecular structure, the solution environment, including the pH, temperature, and ionic strength, organic solvents, and enzymes [107,108]. In this part, we will introduce and discuss the potential strategies for promoting the self-assembly of various biomolecules.

3.1. pH Effect

The growth of biological nanostructures is also influenced by the solution conditions, such as pH, temperature, and ionic strength [109–111]. For instance, the peptide sequence KLVFFAE from the Aβ (16–22) peptide of Alzheimer's (AD) is very sensitive to the environmental pH. Hsieh and co-workers demonstrated that the Aβ (16–22) peptide (KLVFFAE) could self-assemble in neutral and acidic conditions to different nanostructures [112]. Under a neutral condition, the peptide assembled into nanofibers due to the cross-strand pairing between the positively charged K_{16} and the deprotonated C-terminal E_{22} side chain. However, under an acidic pH condition, the peptide formed nanotubes because the protonated E_{22} side chain weakened the K_{16}–E_{22} salt bridge, and the strands shifted out of register and grew into nanotubes (Figure 7a). In another example, Ghosh et al. developed a strategy for precisely tuning the self-assembly behavior of PA by adjusting the solution pH [113]. They found that PA could self-assemble into nanofibers under pH 4 and spherical nano-micelles at pH 10 (Figure 7b). Furthermore, Chen et al. designed a pH-controlled system that could control the PA self-assembly into micelles, nanofibers, and nanofiber bundles due to the clever designs of complementary electrostatic attraction using oppositely charged amino acid pairs, such as arginine and aspartic acid, when the pH was changed (Figure 7c) [114].

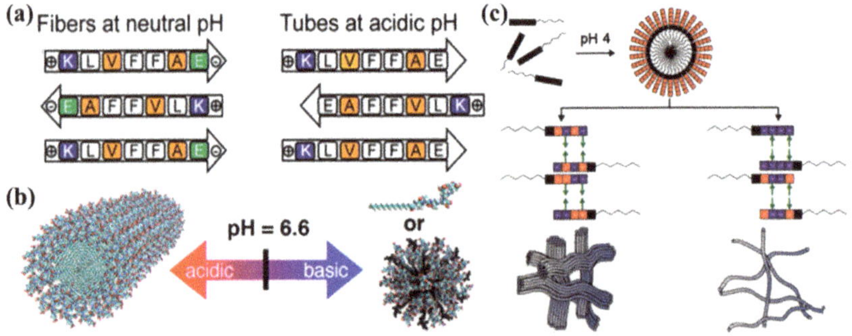

Figure 7. The pH effect on the self-assembly of biomolecules: (**a**) Self-assembled fibers and tubes under different pH conditions. Reprinted with permission from [112]. Copyright 2017 American Chemical Society. (**b**) The pH-triggered morphological transition of self-assembling PA. Reprinted with permission from [113]. Copyright 2012 American Chemical Society. (**c**) Self-assembled nanofibers by pH-mediated lateral assembly. Reprinted with permission from [114]. Copyright 2015 American Chemical Society.

The self-assembly of proteins, enzymes, and viruses is also affected by the solution property [115,116]. For instance, Brodin et al. utilized a Rosetta-interface-designed cytochrome 3 (RIDC3) to self-assemble 1D protein nanotubes and 2D protein nanosheets through Zn^{2+}-coordination [117,118]. In their studies, the morphology of the self-assembled RIDC3 was dependent on the concentration of Zn^{2+}, the RIDC3 concentration, and the pH of the solution. The Zn^{2+}-coordination was reduced when the pH and ([Zn]:[RIDC3]) ratios were decreased to lower conditions; thus, the formed 1D microtubes were transferred into 2D nanosheets due to the reduced nucleation efficiency of Zn-mediated RIDC3 [117]. In addition, 2D Zn^{2+}-RIDC3 arrays were formed under the condition of a lower concentration of Zn^{2+} and a lower pH [118].

The self-assembly of DNA molecules to ordered nanostructures can also be mediated by the pH-responsive formation of a triplex/tetraplex [119–121]. For instance, Wu and Willner recently reported the pH-stimulated reconfiguration and structural isomerization of a DNA origami dimer and trimer by designing pH-sensitive origami dimers and trimers [121]. It was known that triplex DNA nanostructures containing Hoogsteen-type C-G·C$^+$ bridges can be stabilized under acidic conditions and then separated under neutral systems, while the triplex strands, including T-A·T bridges, can be stabilized at neutral pH and then separated under basic systems. On this basis, they performed the pH-stimulated cyclic assembly and separation of the oligomeric origami, and proved the programmed site-specific cleavage of trimer–origami and the reassembly of the separated units. In addition, the pH-mediated isomerization of a linear three-frame origami into a bent configuration has also been proved.

3.2. Temperature Effect

It is well-known that temperature is another important factor that affects the conformation and the intermolecular interactions of biomolecules in solution [122,123].

Previously, Hamley's group found that the PA palmitoyl-KTTKS showed a thermal transition from nanotapes to micelles when the temperature was changed from 20 to 30 °C [124]. In a further study, they studied the effect of temperature on the self-assembly mechanism of PA (C16-KKFFVLK) [125]. They used cryogenic transmission electron microscopy (cryo-TEM), small-angle X-ray and scattering, and circular dichroism (CD) spectra to observe the reversible thermal transition and self-assembly of PA. It was found that PA self-assembled into nanotubes and helical ribbons at room temperature. Interestingly, PA self-assembled into twisted tapes under heating, but the nanotubes and ribbons were reformed under cooling, as shown in Figure 8a.

The self-assembly of polymers and proteins by controlling the temperature has also been studied [101–104]. For instance, a protein–polymer biohybrid was designed by controlling both temperature and pH [126]. Firstly, a hydrophilic initiator (2-bromoisobuta-noic acid N-hydroxysuccinimide (NHS-BiB)) was immobilized onto the surface of an Amelogenin (AME) nanosphere to form a macroinitiator (AME initiator), and then poly(N-isopropylacrylamide) (PNIPAm) chains were grafted to form AME–PNIPAm bioconjugates by temperature-induced self-assembly (Figure 8b). When the temperature was increased from 20 to 40 °C, the hydrodynamic particle size was increased to 218.7 nm with a very narrow size distribution. Huang et al. designed a "rod-coil" graft copolymer containing a polyphenylene backbone linked with poly(ethylene oxide) (PEO) side chains [127], which could form nanoribbons and multilayer sheets at different temperatures.

In another example, the effect of incubation temperature on the self-assembly of regenerated silk fibroin (RSF) was investigated by Zhong and co-workers [128]. They found that the effect of temperature on the self-assembly of RSF was dependent on the concentration of RSF. For a relatively low concentration of RSF, the increase in incubation temperature promoted the formation of anti-parallel β-sheet protofibrils and inhibited the growth of random coil protofilaments/globule-like molecules. However, under a higher concentration of RSF, the increase in incubation temperature changed the morphologies of RSF from protofilaments to protofibrils and beads, and then to longer nanofibers and globules. This work makes it clear that the conformation and morphology of biomolecules can be tuned by controlling the incubation temperature, which will be helpful for us to understand the formation mechanism of various RSF-based biomaterials and extend their biomedical applications.

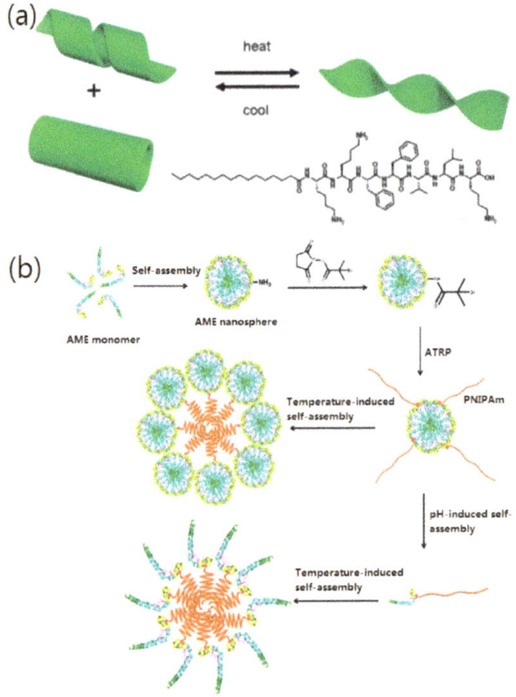

Figure 8. The temperature effect on biomolecular self-assembly: (**a**) The thermo-reversible transition and (bottom right) structure of PA. Reprinted with permission from [125]. Copyright 2013 The Royal Society of Chemistry. (**b**) The synthesis and proposed model of self-assembly and disassembly of pH- and temperature-responsive Amelogenin (AME)–PNIPAm bioconjugates. Reprinted with permission from [126]. Copyright 2018 WILEY-VCH.

3.3. *Ionic Effect*

Along with the effects of pH and temperature, the effect of ions/ionic strength on the self-assembly of proteins [129,130], peptides [131], DNA [132], and RNA [133] molecules have been reported.

Semerdzhiev et al. investigated the self-assembly of α-synuclein into protein fibrils and suprafibrillar by adjusting some external stimulations, including pH, temperature, ionic strength, protein concentration, and seeding [130]. Their results indicated that the formation of suprafibrillar protein assemblies requires a high salt concentration (>10^4 μM K^+/Na^+). However, at a low ionic strength (about 10^2 μM K^+/Na^+), the creation of individual protein fibrils was dominant in the solution due to the strong interfibril repulsion. With the increasing of salt concentration, the electrostatic effects between protein fibrils were screened, promoting the interactions between the formed fibrils and the formation of a sheet-like structure. Further increasing the ionic strength even caused the formation of cylindrical protein aggregates.

Dai and co-workers reported the tunable assembly of an amyloid-forming peptide towards nanosheet structures [131]. They found that the size and yield of the self-assembled amyloid peptide (KLVFFAK) nanosheets could be fine-tuned by adjusting the ionic strength in aqueous solution. With the increasing of NaCl concentration from 0.1 to 1.0 M, the width and the yield of self-assembled peptide nanosheets increased accordingly. They suggested that salt could improve the aggregation ability of peptide molecules by screening out the repulsive interactions between the positively charged Lys-Lys contacts.

The self-assembly of DNA and RNA is also affected by the ions and ionic strength. For instance, Liu and co-workers demonstrated the self-assembly of DNA on a mica surface [132], which can be regulated by changing the concentration of Ni^{2+} to form a salt bridge between DNA and the mica surface. They found that a suitable Ni^{2+} concentration was crucial for the formation of 2D DNA arrays on the mica surface. A low Ni^{2+} concentration did not provide enough attractive force to bind DNA to the mica surface, while an Ni^{2+} concentration that was too high caused a strong DNA–surface attraction and hindered the DNA mobility and self-assembly. AFM experiments indicated that 2D DNA trihexagonal (Figure 9a), square (Figure 9b), and rhombic (Figure 9c) arrays were formed by controlling the self-assembly of the four-pointed-star DNA motif via adding 3, 4, and 6 mM Ni^{2+}, respectively. This study proved that the weak DNA–DNA interactions could be stabilized by using a suitable ionic concentration to regulate the DNA–surface interactions, promoting the formation of larger DNA nanostructures. Recently, Yang and co-workers demonstrated a novel K^+ ion-stimulated self-assembly of DNA origami nanostructures by using G-quadruplexes as stimuli-responsive bridges [134]. It was found that, with the stimulation of monovalent cations, the conformation transitions between the G-quadruplex and its sing-strand state promoted the reversible assembly process of DNA origami. Their study provides a potential strategy for designing pH-responsive DNA nanomaterials for biomedical and nanotechnological applications. In another study, Garmann et al. studied the assembly pathway of an icosahedral ssRNA virus and found that the in-vitro assembly of ssRNA virus was affected by the pH and ionic strength [133].

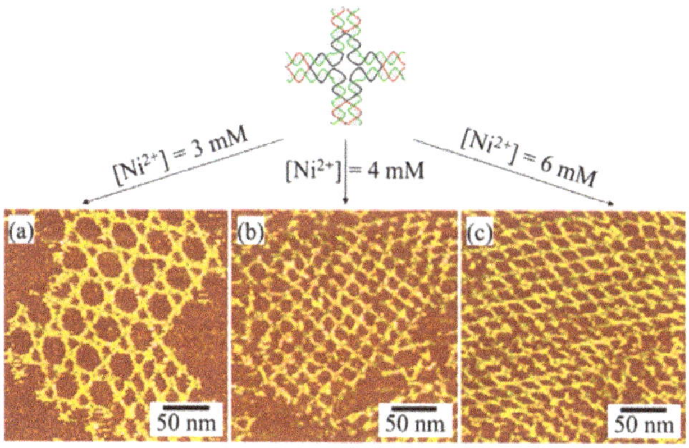

Figure 9. Self-assembled DNA nanostructures on a mica surface by adjusting the Ni^{2+} ion concentration: (**a**) 3, (**b**) 4, and (**c**) 6 mM. Reprinted with permission from [132]. Copyright 2017 Wiley-VCH.

3.4. Organic Stimulators

It is known that the structural formation of self-assembled biological nanomaterials is related to the intermolecular noncovalent interactions, including hydrogen bonds, electrostatic interactions, hydrophobic interactions, and π–π interactions. However, the conformation transition and self-assembly of biomolecules are affected by some organic solvents due to the synergistic effects with these interactions.

The self-assembly of peptides and proteins into various morphologies in different organic solvents have been investigated widely [135–141]. For instance, Yan et al. investigated the self-assembly of a small FF peptide in chloroform and toluene, and found that the peptide can self-assemble into long nanofibrils and then entangle to form organogels [135]. It was found that the created FF-based organogels were thermo-responsive and the sol-gel process was thermo-reversible. In a further study, they investigated the effects of organic co-solvents (ethanol and toluene) on the stabilization of the created organogels [136]. Flower-like microcrystals were prepared by a further self-assembly process of gels (Figure 10a). The solvent (ethanol) has a higher polarity than toluene, and, therefore, it caused the formation of hydrogen bonds during biomolecular self-assembly and promoted the subsequent transition of organgels to microcrystals.

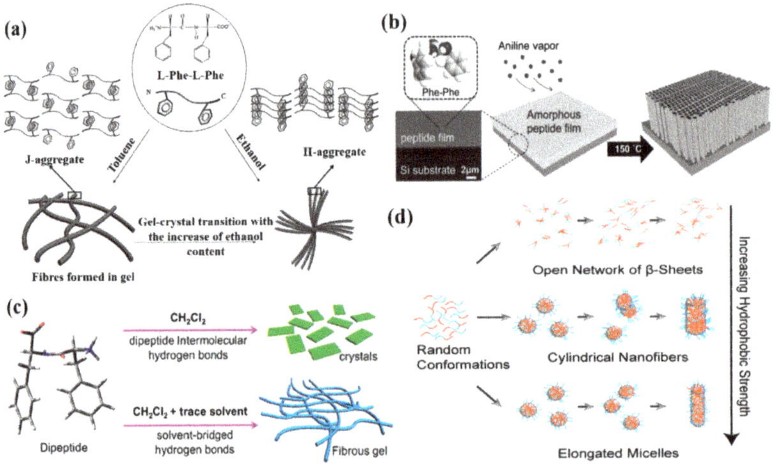

Figure 10. Effects of organic solvents on biomolecular self-assembly: (**a**) The structural transition of FF nanofibers in mixed organic solvents. Reprinted with permission from [136]. Copyright 2010 WILEY-VCH. (**b**) Vertically well-aligned peptide nanowires prepared by high-temperature aniline vapor aging. Reprinted with permission from [140]. Copyright 2008 WILEY-VCH. (**c**) A phase transition induced by trace amounts of organic solvent. Reprinted with permission from [142]. Copyright 2016 American Chemical Society. (**d**) Kinetic mechanisms of peptide self-assembly studied by molecular dynamics simulation (MDS). Reprinted with permission from [143]. Copyright 2015 American Chemical Society.

Ryu et al. investigated the high-temperature-induced self-assembly of an FF peptide into vertically aligned nanowires in the environment of aniline vapor [140]. In their work, an FF peptide solution dissolved in 1,1,1,3,3,3-Hexafluoro-2-propanol (HFIP) was first dropped onto a silicon wafer or quartz plate substrate. Subsequently, the peptide solution was dried in a vacuum desiccator to form a patterned FF film. Finally, the FF film was aged under an aniline vapor condition at 150 °C to obtain vertical peptide nanowire arrays on a silicon substrate, as shown in Figure 10b.

Recently, Wang et al. found that a trace amount of solvent can be a predominant factor to control the self-assembly of an FF peptide in dichloromethane, ethanol, *N*,*N*-dimethylformamide (DMF), and acetone [142]. They demonstrated that hydrogen bonding plays more of a role in the process of nanofiber formation than other noncovalent interactions (Figure 10c), and that the bonding of C=O and N–H in FF molecules was affected by the used organic solvents.

To further understand the effects of organic solvents on the self-assembly of peptides, Fu et al. used molecular dynamics simulations (MDSs) to study the solvent effects on the self-assembly of PAs [143]. When the hydrophobic interaction was weak, biomolecular aggregates were formed due to the hydrophobic interaction and hydrogen bonding. In addition, the aggregates could grow with different directions, resulting in the formation of an open network structure. However, the structure was changed from open one to a closed one when the hydrophobic interaction was increased. When the hydrophobic interaction was further increased, the hydrogen bonds were reduced and all of the peptides appeared in a random-coil conformation and formed an elongated micelle structure, as indicated in Figure 10d.

The above examples and discussions provide experimental and theoretical evidence that organic solvents exert significant effects on the conformation transition and self-assembly pathways of biomolecules.

3.5. Enzymatic Stimulators

Enzymes can also influence the self-assembly of biomolecules significantly as they may catalyze the formation of biological materials. Based on the functions and types, enzymes can promote or inhibit the aggregation and self-assembly of biomolecules [144]. In recent decades, many studies on enzymatic driving for the formation of NPs, nanofibrils, crosslinked hydrogels, and other superstructures have been reported [145,146].

For example, Amir et al. introduced a novel enzyme-triggered strategy to mediate the self-assembly of a block coplymer into NPs under a physiological condition [147]. In their work, they designed a water-soluble diblock copolymer containing a hydrophilic block copolymer and a block of phosphorylated 4-hydroxystyrene, and then the phosphate groups of the copolymer were removed by phosphatase to form the amphiphilic diblock copolymers. Subsequently, the amphiphilic copolymers were self-assembled into colloidal NPs via an in-situ process (Figure 11a). This approach constitutes a new way to form polymeric materials by using various polymeric backbones and enzymatic triggers.

Besides biopolymers, the self-assembly of other biomolecules, such as proteins, peptides, and DNA, could be driven by enzymes [148–151]. Previously, Xu's group used the enzyme-instructed self-assembly (EISA) approach to prepared a few supramolecular nanostructures, such as nanofibers and hydrogels [13,152–154]. For example, they designed a series of structural precursors based on the peptide GNNQQNY sequence of the yeast prion Sup35, which can self-assemble to form supramolecular hydrogels induced by alkaline phosphatase in water (Figure 11b) [152]. In another study [155], they investigated the enzyme-induced in-situ self-assembly of C-terminal methylated phosphotetrapeptide (pTP-Me) into PNFs. It was found that the obtained PNFs exhibited strong synergism with NF-κB targeting for the selective necroptosis of cancer cells (Figure 11c).

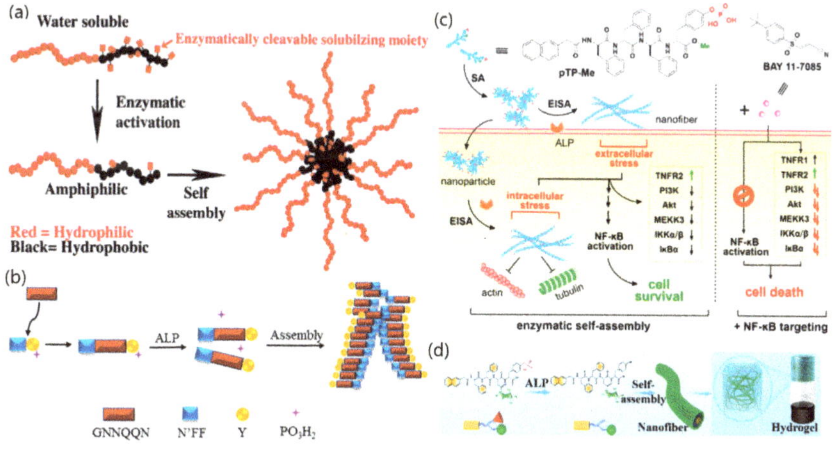

Figure 11. The enzyme-mediated self-assembly of biomolecules: (**a**) The enzyme-active self-assembly of water-soluble diblock copolymers to colloidal nanostructures. Reprinted with permission from [147]. Copyright 2009 American Chemical Society. (**b**) Alkaline phosphatase (ALP)-mediated formation of a peptide hydrogel. Reprinted with permission from [152]. Copyright 2016 The Royal Society of Chemistry. (**c**) The enzyme-induced self-assembly of pTP-Me into PNFs. Reprinted with permission from [155]. Copyright 2018 American Chemical Society. (**d**) The self-assembly of a glycopeptide to a supramolecular hydrogel. Reprinted with permission from [156]. Copyright 2018 American Chemical Society.

Qi and co-workers developed a novel hydrogel from the enzyme-induced supramolecular self-assembly of a synthetic glycopeptide to mimic the glycosylated microenvironment of the extracellular matrix [156]. In their work, a gelator precursor 1 glycopeptide, containing a naphthyl group, a tetrapeptide motif (Phe-Phe-Asp-Tyr(H_2PO_3)), and a sugar moiety (D-glucosamine), was first prepared via a solid-phase synthesis method. Then, the glycopeptide was dissolved in water with a pH value of 7.4 and transferred into gelator 2 by adding alkaline phosphatase (owing to the enzymatic dephosphorylation towards gelator 1). After that, the gelator self-assembled into nanofibers and then into a hydrogel at room temperature via aromatic–aromatic and hydrogen bonding interactions, as shown in Figure 11d. Furthermore, the fabricated hydrogel could serve as biomimetic scaffold to promote the generation of new blood capillaries in vitro and vivo.

3.6. Photo-Stimulation

Various biological nanostructures can also be obtained by the photo-induced self-assembly of biomolecules, such as peptides [157,158] and DNA [159–161]. The photo-triggered assembly of biomolecules exhibits a few advantages, such as reversibility, rapidity, remoteness, and cleanliness. In the photo-triggered assembly process, the photo-responsive groups acted as photoswitching units to mediate the structure and functions of the formed nanostructures.

Previously, Muraoka et al. synthesized photo-responsive PAs with a palmitoyl tail, the 2-nitrobenzyl group, and an oligopeptide motif ($GV_3A_3E_3$), which were capable of self-assembling into supramolecular quadruple nanofibers [157]. Under the irradiation of light at 350 nm, the 2-nitrobenzyl group was cleaved, which dissociates the quadruple helical fibers to single non-helical fibrils. Ma and co-workers designed a photoswitchable molecule that can co-assemble with a cationic FF peptide to form elongated nanoplates and helical nanobelts under visible light [158]. After UV irradiation, the photo-isomerization of the photoswitchable molecule caused the disassembly of peptides to vesicle-like structures.

Similar to peptides, DNA molecules can also be induced by light stimulation to form self-assembled nanostructures. For instance, Tanaka and co-workers demonstrated the robust and photo-controllable synthesis of DNA structures (three-point-star motifs and capsules) by UV irradiation to the azobenzenes that were inserted into the sticky ends of DNA motifs [159]. Without the UV irradiation, the three-point-star motifs with azobenzenes self-assembled to sphere-shaped capsules, which were broken down into three-point-star motifs after UV irradiation for 50 s. Their study enhanced the potential of self-assembled DNA nanomaterials for controllable biomedical applications, such as precise drug delivery. Sugiyama and co-workers presented the fabrication of predesigned multiorientational patterns by photo-induced self-assembly of DNA origami nanostructures [160]. Firstly, they designed a series of 50-nm-sized hexagonal DNA origamis, which were then functionalized with photo-responsive oligonucleotides. Under visible light irradiation, the DNA origami self-assembled into predesigned oligomeric nanostructures, which could then disassemble into DNA origami structures under optimal UV irradiation at 40 °C. In a further study, they investigated the in-situ dynamic assembly/disassembly processes of photo-responsive DNA origami nanostructures, which can be placed on a lipid membrane surface [161]. It was found that the bilayer-placed DNA hexagonal structure was disassembled into monomers under UV irradiation, and reassembled into a larger DNA dimer after visible light irradiation.

All of the above cases prove the feasibility of photo-stimulation in the control of biomolecular self-assembly and the formation of various nanostructures.

3.7. Tailoring Molecular Structure

Molecular structure is crucial for guiding the self-assembly of biomolecules (especially for peptide and DNA molecules) into well-ordered superstructures [162,163]. For instance, by designing peptide sequences with multiple functions, such as recognition, binding, signal acceptor, and self-assembly motifs, it is easy to create 1D, 2D, and 3D peptide superstructures with tailored functions [8].

Meanwhile, it is possible to synthesize DNA superstructures by designing DNA sequences and other complex DNA building blocks [163].

Dai et al. used an amyloid peptide to self-assemble 2D peptide nanosheets (PNSs) by adjusting the molecular structure [131]. They changed the peptide sequence KLVFFAK into KLVFGAK and VQIVAK to provide the possibility of a β-sheet along the zippering axis face-to-face with the back pattern, and the three peptide sequences (KLVFFAK, KLVFGAK, and VQIVAK) could self-assemble into 2D nanosheets and nanofibers, respectively. Hence, the molecular structure of the peptide is very important for self-assembly into a well-ordered structure. In another case, Sun et al. discussed the self-assembly behaviors of three designed RADA16-1 peptides by studying the effects of motifs, pH, and assembly time [162]. Three functional peptide motifs, IKVAV, RGD, and YIGSR, were utilized to modify the RADA 16-1 peptide to provide different net charges and amphiphilic properties of the designed peptides at neutral pH. The obtained results indicated that both the electrostatic and hydrophilic/hydrophobic interactions of the motifs affected the self-assembly of the peptide and the morphologies of the formed PNFs.

Wei et al. designed several peptides with various functional motifs for the creation of functional 1D PNFs towards biomineralization, sensors, and cell targeting [33,76,77]. Very recently, they designed a novel peptide sequence (LLVFGAKMLPHHGA) to create 2D functional PNSs, as shown in Figure 12 [164]. The results indicated that the motif of LLVFGAK was responsible for the self-assembly and formation of PNSs, and KMLPHHGA provided an additional function for the biomineralization of HA. Therefore, the designed bifunctional PNSs exhibited unique properties for binding with 3D graphene foam (GF) to fabricate 3D biominerals.

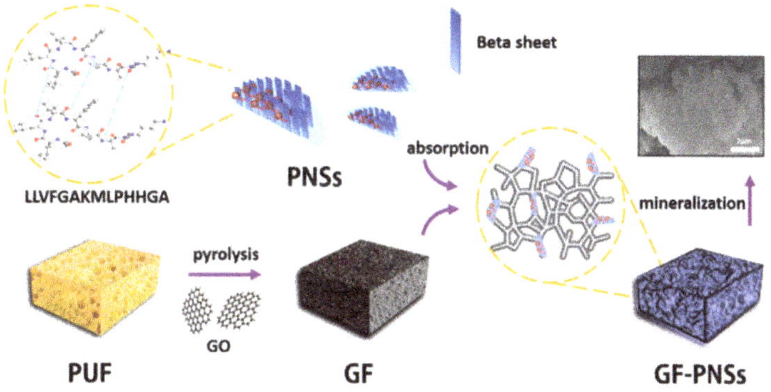

Figure 12. Two-dimensional peptide nanosheets (PNSs) by molecular tailoring: a schematic presentation of 2D peptide self-assembly and the biomimetic fabrication of 3D graphene foam (GF)-PNS-HA minerals. Reprinted with permission from [164]. Copyright 2018 WILEY-VCH.

Here, it is highly recommended for the authors to read two recent review papers on the design of small bioactive [165] and protein-mimic peptides [9] for biomaterials design and biomedical applications.

4. Various Self-Assembled Biological Nanostructures/Materials

Based on the above discussion, it can be concluded that hydrogen bonds, electrostatic interactions, hydrophobic interactions, and π–π interactions play important roles in mediating the self-assembly of biomolecules and promoting the formation of biological nanostructures. Other factors, such as molecular structure, pH effect, temperature, organic stimulators, and enzymatic stimulators are crucial for the self-assembly of biomolecules. To make it more clear, we summarize the types, nanostructures, interactions, and effect factors of the self-assembly of various biomolecules in Table 1.

Table 1. A summary of the formed nanostructures via biomolecular self-assembly, and the internal interactions as well as external stimulations.

Biomolecules	Nanostructures	Interactions	Stimulations	Ref.
Proteins				
SP1	Nanowires	Electrostatic	Micelles	[60]
SP1	Nanowire-QDs	Electrostatic	Enzyme	[61]
BSA	NPs	Hydrophobic	Organic	[72]
IgG	2D crystals	Ligand–receptor	-	[99]
RIDC3	Nanotubes/2D Crystals	Zn^{2+}-coordination	pH	[117,118]
Amelogenin	Nanospheres	-	pH and temperature	[126]
Silk fibroin	Protofibrils/Fibers	-	temperature	[128]
A-synuclein	Fibrils	Electrostatic	Ions	[130]
Peptides				
FF	Fibers/Tubes/Rods	Hydrogen bonds	Organic	[21]
FF	PNWs-G	Hydrogen bonds and $\pi-\pi$ interaction	Organic	[52]
VIAGASLWWSEKLVIA	GN-PNF-AgNW	Electrostatic	Ethanol	[58]
NapFFKYp	Nanofibers	Hydrophobic	Organic	[65]
EAK 16-II	Nanofibers	Electrostatic/hydrophobic	Molecular structure	[66]
RGDAEAKAEAKYWYAFAEAKAEAKRGD	PNF-GQDs	π–π/Electrostatic	ethanol	[76]
AEAKAEAKYWYAFAEAKAEAK	GO-PNF	π–π/Electrostatic	Ethanol	[33]
AEAKAEAKYWYAFAEAKAEAK	GQD-PNF-GO	π–π/Electrostatic	Ethanol	[77]
Peptide	Fibers/Aggregates	Ligand–receptor	Enzyme	[96]
KLVFFAE	Nanofibers/Tubes	Electrostatic	pH	[112]
PA	Micelles/Nanofibers	Electrostatic	pH	[114]
C_{16}-KKFFVLK	Nanotubes/Helical ribbons	Hydrogen bonds	Temperature	[125]
KLVFFAK	Nanosheets	Electrostatic	Ionic strength	[131]
GNNQQNY	Hydrogels	Hydrogen bonds	Enzyme	[152]
FFDY(H_2PO_3)	Fibers/Hydrogels	π–π/Hydrogen bonds	Enzyme	[156]
GV3A3E3	Fibers	Hydrogen bonds/hydrophobic	Light	[157]
FF	Nanoplates/belts	Hydrogen bonds/π–π	Light	[158]
DNA/RNA				
DNA	GQDs-ionic liquid (IL)-NF-DNA	π–π interactions	Enzyme	[81]
DNA	GO-DNA	π–π interactions	Temperature	[82]
DNA	Hydrogels	Clamped hybridization	-	[24]
DNA	2D lattices	base pairing	Buffer/Mg^{2+}	[83]
DNA	Tiles	base pairing	Mg^{2+}	[84]
DNA	Nanowires/Sheets	base pairing	-	[85]
DNA	2D arrays	base pairing	Ni^{2+}	[132]
DNA	Capsules	base pairing	Light	[159]
DNA	Origami	base pairing	Light	[160]
DNA	Origami	base pairing	Light	[161]
RNA	Tetrahedrons	RNA packing	-	[91]
RNA	Triangles	RNA packing	-	[92]
RNA	Lattices/Tubes	RNA packing	-	[93]
PNA	Fibers	π–π and base pairing	-	[94]
Virus				
CCMV	3D crystals	Ligand-receptor	-	[95]
Bacteriophage P22	P22VLP-NPs	Electrostatic interaction	NPs	[59]
Enzymes				
OxOx/HRP	CRGO-enzyme	Hydrophobic	pH	[73]
GOx/CAT	graphene nanodots-porous gold	π-π	Organic	[80]
Other biopolymers				
cholesterol	Microrods/ribbons	-	Polymer	[104]
cholesterol	Aggregates	-	Polymer	[105]

CRGO, chemically reduced graphene oxide.

5. Conclusions and Outlooks

The self-assembly of biomolecules provides a direct and effective way to create functional nanostructures and nanomaterials. Our deep understanding of the self-assembly mechanisms of biomolecules makes it possible to design and synthesize many novel biological nanomaterials with specific functions. In this review, we demonstrated the self-assembly of biomolecules into pure and hybrid biological nanomaterials from two perspectives (internal interactions and external stimulations) by introducing and discussing relevant cases. This work will be helpful for readers to understand basic methods to promote the self-assembly of biomolecules, develop novel biological nanomaterials, and explore the potential applications of self-assembled biological nanomaterials in materials science, biomedical engineering, tissue engineering, analytical science, and the energy and environmental sciences.

Biomolecular self-assembly towards functional nanomaterials has been one of the most focused-on fields in the last few years. In our opinion, the further development in this research field may include the following. First, the design of functional motifs for creating functional nanomaterials via self-assembly could be further studied. For example, the design of peptide molecules by combining a few functional motifs will form 1D to 3D nanostructures with multiple functions via peptide self-assembly. The design of DNA motifs can create uniform DNA nanostructures from nanowires to nanosheets and microcrystals through DNA hybridization. Second, it is important to develop bioinspired synthesis strategies by using self-assembled biological nanostructures to fabricate functional hybrid nanomaterials [166]. For instance, the conjugation between biological nanostructures and bioinspired NPs, QDs, and biominerals (such as HA and $CaCO_3$) could introduce new properties and functions to the designed hybrid nanomaterials. Third, extensions to the applications of the self-assembled and bioinspired nanomaterials should be explored. More attention should be paid to the fabrication of energy storage materials and environment-related materials or techniques (such as filters, membranes, and sensing techniques).

Author Contributions: L.W. and G.W. proposed the project. All of the authors carried out the search for reference materials and the data analysis. L.W., C.G., X.Y., and G.W. wrote the manuscript. G.W. performed a review and the final edit.

Acknowledgments: L.W., X.Y., and G.W. acknowledge the financial support from the National Natural Science Foundation of China (Grant No. 21505049). C.G. and G.W. thank the National Natural Science Foundation of China (No. 51873225), the Chinese Scholarship Council (CSC), and the Deutsche Forschungsgemeinschaft (No. WE 5837/1-1) for the financial support.

Conflicts of Interest: The authors declare no conflict of interest.

References

1. Gwo, S.; Chen, H.Y.; Lin, M.H.; Sun, L.Y.; Li, X.Q. Nanomanipulation and controlled self-assembly of metal nanoparticles and nanocrystals for plasmonics. *Chem. Soc. Rev.* **2016**, *45*, 5672–5716. [CrossRef] [PubMed]
2. Wang, L.; Sun, Y.J.; Li, Z.; Wu, A.G.; Wei, G. Bottom-up synthesis and sensor applications of biomimetic nanostructures. *Materials* **2016**, *9*, 53. [CrossRef]
3. Hoheisel, T.N.; Hur, K.; Wiesner, U.B. Block copolymer-nanoparticle hybrid self-assembly preface. *Prog. Polym. Sci.* **2015**, *40*, 3–32. [CrossRef]
4. Bhattacharyya, K.; Mukherjee, S. Fluorescent metal nano-clusters as next generation fluorescent probes for cell imaging and drug delivery. *Bull. Chem. Soc. Jpn.* **2018**, *91*, 447–454. [CrossRef]
5. Komiyama, M.; Mori, T.; Ariga, K. Molecular imprinting: Materials nanoarchitectonics with molecular information. *Bull. Chem. Soc. Jpn.* **2018**, *91*, 1075–1111. [CrossRef]
6. Rogers, W.B.; Shih, W.M.; Manoharan, V.N. Using DNA to program the self-assembly of colloidal nanoparticles and microparticles. *Nat. Rev. Mater.* **2016**, *1*, 16008. [CrossRef]
7. Bai, Y.S.; Luo, Q.; Liu, J.Q. Protein self-assembly via supramolecular strategies. *Chem. Soc. Rev.* **2016**, *45*, 2756–2767. [CrossRef] [PubMed]
8. Wei, G.; Su, Z.Q.; Reynolds, N.P.; Arosio, P.; Hamley, I.W.; Gazit, E.; Mezzenga, R. Self-assembling peptide and protein amyloids: From structure to tailored function in nanotechnology. *Chem. Soc. Rev.* **2017**, *46*, 4661–4708. [CrossRef]
9. Zhang, W.S.; Yu, X.Q.; Li, Y.; Su, Z.Q.; Jandt, K.D.; Wei, G. Protein-mimetic peptide nanofibers: Motif design, self-assembly synthesis, and sequence-specific biomedical applications. *Prog. Polym. Sci.* **2018**, *80*, 94–124. [CrossRef]

10. Milles, S.; Jensen, M.R.; Communie, G.; Maurin, D.; Schoehn, G.; Ruigrok, R.W.H.; Blackledge, M. Self-assembly of measles virus nucleocapsid-like particles: Kinetics and rna sequence dependence. *Angew. Chem. Int. Ed.* **2016**, *55*, 9356–9360. [CrossRef]
11. Sawada, T.; Serizawa, T. Filamentous viruses as building blocks for hierarchical self-assembly toward functional soft materials. *Bull. Chem. Soc. Jpn.* **2018**, *91*, 455–466. [CrossRef]
12. Zhou, J.; Du, X.W.; Berciu, C.; He, H.J.; Shi, J.F.; Nicastro, D.; Xu, B. Enzyme-instructed self-assembly for spatiotemporal profiling of the activities of alkaline phosphatases on live cells. *Chem* **2016**, *1*, 246–263. [CrossRef] [PubMed]
13. Wang, H.M.; Feng, Z.Q.Q.; Wang, Y.Z.; Zhou, R.; Yang, Z.M.; Xu, B. Integrating enzymatic self-assembly and mitochondria targeting for selectively killing cancer cells without acquired drug resistance. *J. Am. Chem. Soc.* **2016**, *138*, 16046–16055. [CrossRef]
14. Lin, Y.Y.; Chapman, R.; Stevens, M.M. Integrative self-assembly of graphene quantum dots and biopolymers into a versatile biosensing toolkit. *Adv. Funct. Mater.* **2015**, *25*, 3183–3192. [CrossRef] [PubMed]
15. Liu, B.; Cao, Y.Y.; Huang, Z.H.; Duan, Y.Y.; Che, S.N. Silica biomineralization via the self-assembly of helical biomolecules. *Adv. Mater.* **2015**, *27*, 479–497. [CrossRef] [PubMed]
16. Stephanopoulos, N.; Ortony, J.H.; Stupp, S.I. Self-assembly for the synthesis of functional biomaterials. *Acta Mater.* **2013**, *61*, 912–930. [CrossRef] [PubMed]
17. Mauro, M.; Aliprandi, A.; Septiadi, D.; Kehra, N.S.; De Cola, L. When self-assembly meets biology: Luminescent platinum complexes for imaging applications. *Chem. Soc. Rev.* **2014**, *43*, 4144–4166. [CrossRef] [PubMed]
18. Zhu, G.Z.; Hu, R.; Zhao, Z.L.; Chen, Z.; Zhang, X.B.; Tan, W.H. Noncanonical self-assembly of multifunctional DNA nanoflowers for biomedical applications. *J. Am. Chem. Soc.* **2013**, *135*, 16438–16445. [CrossRef]
19. Wang, L.; Wu, A.G.; Wei, G. Graphene-based aptasensors: From molecule-interface interactions to sensor design and biomedical diagnostics. *Analyst* **2018**, *143*, 1526–1543. [CrossRef]
20. Wang, L.; Zhang, Y.J.; Wu, A.G.; Wei, G. Designed graphene-peptide nanocomposites for biosensor applications: A review. *Anal. Chim. Acta* **2017**, *985*, 24–40. [CrossRef]
21. Li, Q.; Jia, Y.; Dai, L.R.; Yang, Y.; Li, J.B. Controlled rod nanostructured assembly of diphenylalanine and their optical waveguide properties. *ACS Nano* **2015**, *9*, 2689–2695. [CrossRef] [PubMed]
22. Caplan, M.R.; Moore, P.N.; Zhang, S.G.; Kamm, R.D.; Lauffenburger, D.A. Self-assembly of a beta-sheet protein governed by relief of electrostatic repulsion relative to van der waals attraction. *Biomacromolecules* **2000**, *1*, 627–631. [CrossRef] [PubMed]
23. Elsawy, M.A.; Smith, A.M.; Hodson, N.; Squires, A.; Miller, A.F.; Saiani, A. Modification of beta-sheet forming peptide hydrophobic face: Effect on self-assembly and gelation. *Langmuir* **2016**, *32*, 4917–4923. [CrossRef] [PubMed]
24. Wang, J.B.; Chao, J.; Liu, H.J.; Su, S.; Wang, L.H.; Huang, W.; Willner, I.; Fan, C.H. Clamped hybridization chain reactions for the self-assembly of patterned DNA hydrogels. *Angew. Chem. Int. Ed.* **2017**, *56*, 2171–2175. [CrossRef] [PubMed]
25. Wei, G.; Reichert, J.; Bossert, J.; Jandt, K.D. Novel biopolymeric template for the nucleation and growth of hydroxyapatite crystals based on self-assembled fibrinogen fibrils. *Biomacromolecules* **2008**, *9*, 3258–3267. [CrossRef]
26. Wei, G.; Reichert, J.; Jandt, K.D. Controlled self-assembly and templated metallization of fibrinogen nanofibrils. *Chem. Commun.* **2008**, 3903–3905. [CrossRef]
27. Dave, A.C.; Loveday, S.M.; Anema, S.G.; Jameson, G.B.; Singh, H. Modulating beta-lactoglobulin nanofibril self-assembly at pH 2 using glycerol and sorbitol. *Biomacromolecules* **2014**, *15*, 95–103. [CrossRef]
28. Li, R.; Horgan, C.C.; Long, B.; Rodriguez, A.L.; Mather, L.; Barrow, C.J.; Nisbet, D.R.; Williams, R.J. Tuning the mechanical and morphological properties of self-assembled peptide hydrogels via control over the gelation mechanism through regulation of ionic strength and the rate of ph change. *RSC Adv.* **2015**, *5*, 301–307. [CrossRef]
29. Kim, W.; Thevenot, J.; Ibarboure, E.; Lecommandoux, S.; Chaikof, E.L. Self-assembly of thermally responsive amphiphilic diblock copolypeptides into spherical micellar nanoparticles. *Angew. Chem. Int. Ed.* **2010**, *49*, 4257–4260. [CrossRef]

30. Yu, X.L.; Du, R.F.; Li, B.Y.; Zhang, Y.H.; Liu, H.J.; Qu, J.H.; An, X.Q. Biomolecule-assisted self-assembly of cds/mos2/graphene hollow spheres as high-efficiency photocatalysts for hydrogen evolution without noble metals. *Appl. Catal. B* **2016**, *182*, 504–512. [CrossRef]
31. Zhao, Z.; Liu, Y.; Yan, H. DNA origami templated self-assembly of discrete length single wall carbon nanotubes. *Org. Biomol. Chem.* **2013**, *11*, 596–598. [CrossRef] [PubMed]
32. Wei, G.; Zhang, Y.; Steckbeck, S.; Su, Z.Q.; Li, Z. Biomimetic graphene-fept nanohybrids with high solubility, ferromagnetism, fluorescence, and enhanced electrocatalytic activity. *J. Mater. Chem.* **2012**, *22*, 17190–17195. [CrossRef]
33. Wang, J.H.; Ouyang, Z.F.; Ren, Z.W.; Li, J.F.; Zhang, P.P.; Wei, G.; Su, Z.Q. Self-assembled peptide nanofibers on graphene oxide as a novel nanohybrid for biomimetic mineralization of hydroxyapatite. *Carbon* **2015**, *89*, 20–30. [CrossRef]
34. Zhang, Q.; Li, M.X.; Zhu, C.Y.; Nurumbetov, G.; Li, Z.D.; Wilson, P.; Kempe, K.; Haddleton, D.M. Well-defined protein/peptide-polymer conjugates by aqueous cu-lrp: Synthesis and controlled self-assembly. *J. Am. Chem. Soc.* **2015**, *137*, 9344–9353. [CrossRef] [PubMed]
35. Zou, Q.L.; Abbas, M.; Zhao, L.Y.; Li, S.K.; Shen, G.Z.; Yan, X.H. Biological photothermal nanodots based on self-assembly of peptide porphyrin conjugates for antitumor therapy. *J. Am. Chem. Soc.* **2017**, *139*, 1921–1927. [CrossRef] [PubMed]
36. Zhang, S.G. Fabrication of novel biomaterials through molecular self-assembly. *Nat. Biotechnol.* **2003**, *21*, 1171–1178. [CrossRef] [PubMed]
37. Groeger, C.; Lutz, K.; Brunner, E. Biomolecular self-assembly and its relevance in silica biomineralization. *Cell Biochem. Biophys.* **2008**, *50*, 23–39. [CrossRef] [PubMed]
38. Zhang, S.G.; Marini, D.M.; Hwang, W.; Santoso, S. Design of nanostructured biological materials through self-assembly of peptides and proteins. *Curr. Opin. Chem. Biol.* **2002**, *6*, 865–871. [CrossRef]
39. Yang, L.L.; Liu, A.J.; Cao, S.Q.; Putri, R.M.; Jonkheijm, P.; Cornelissen, J.J.L.M. Self-assembly of proteins: Towards supramolecular materials. *Chem. Eur. J.* **2016**, *22*, 15570–15582. [CrossRef]
40. Willner, I.; Willner, B. Biomolecule-based nanomaterials and nanostructures. *Nano Lett.* **2010**, *10*, 3805–3815. [CrossRef]
41. McManus, J.J.; Charbonneau, P.; Zaccarelli, E.; Asherie, N. The physics of protein self-assembly. *Curr. Opin. Colloid Interface Sci.* **2016**, *22*, 73–79. [CrossRef]
42. Qi, G.B.; Gao, Y.J.; Wang, L.; Wang, H. Self-assembled peptide-based nanomaterials for biomedical imaging and therapy. *Adv. Mater.* **2018**, *30*, 1703444. [CrossRef]
43. Marchesan, S.; Vargiu, A.V.; Styan, K.E. The phe-phe motif for peptide self-assembly in nanomedicine. *Molecules* **2015**, *20*, 19775–19788. [CrossRef] [PubMed]
44. Lombardo, D.; Kiselev, M.A.; Magazu, S.; Calandra, P. Amphiphiles self-assembly: Basic concepts and future perspectives of supramolecular approaches. *Adv. Cond. Matter. Phys.* **2015**, *2015*, 151683. [CrossRef]
45. Wang, Z.G.; Ding, B.Q. Engineering DNA self-assemblies as templates for functional nanostructures. *Acc. Chem. Res.* **2014**, *47*, 1654–1662. [CrossRef] [PubMed]
46. Shaham-Niv, S.; Adler-Abramovich, L.; Schnaider, L.; Gazit, E. Extension of the generic amyloid hypothesis to nonproteinaceous metabolite assemblies. *Sci. Adv.* **2015**, *1*, e1500137. [CrossRef]
47. Shaham-Niv, S.; Arnon, Z.A.; Sade, D.; Lichtenstein, A.; Shirshin, E.A.; Kolusheva, S.; Gazit, E. Intrinsic fluorescence of metabolite amyloids allows label-free monitoring of their formation and dynamics in live cell. *Angew. Chem. Int. Ed.* **2018**, *57*, 12444–12447. [CrossRef]
48. Shaham-Niv, S.; Rehak, P.; Zaguri, D.; Kolusheva, S.; Kral, P.; Gazit, E. Metabolite amyloid-like fibrils interact with model membranes. *Chem. Commun.* **2018**, *54*, 4561–4564. [CrossRef] [PubMed]
49. Bartocci, S.; Berrocal, J.A.; Guarracino, P.; Grillaud, M.; Franco, L.; Mba, M. Peptide-driven charge-transfer organogels built from synergetic hydrogen bonding and pyrene-naphthalenediimide donor-acceptor interactions. *Chem. Eur. J.* **2018**, *24*, 2920–2928. [CrossRef] [PubMed]
50. Wu, Y.L.; Bobbitt, N.S.; Logsdon, J.L.; Powers-Riggs, N.E.; Nelson, J.N.; Liu, X.L.; Wang, T.C.; Snurr, R.Q.; Hupp, J.T.; Farha, O.K.; et al. Tunable crystallinity and charge transfer in two-dimensional g-quadruplex organic frameworks. *Angew. Chem. Int. Ed.* **2018**, *57*, 3985–3989. [CrossRef]
51. Bilbao, N.; Destoop, I.; De Feyter, S.; Gonzalez-Rodriguez, D. Two-dimensional nanoporous networks formed by liquid-to-solid transfer of hydrogen-bonded macrocycles built from DNA bases. *Angew. Chem. Int. Ed.* **2016**, *55*, 659–663. [CrossRef]

52. Li, P.P.; Chen, X.; Yang, W.S. Graphene-induced self-assembly of peptides into macroscopic-scale organized nanowire arrays for electrochemical nadh sensing. *Langmuir* **2013**, *29*, 8629–8635. [CrossRef] [PubMed]
53. Lee, Y.M.; Jung, B.; Kim, Y.H.; Park, A.R.; Han, S.; Choe, W.S.; Yoo, P.J. Nanomesh-structured ultrathin membranes harnessing the unidirectional alignment of viruses on a graphene-oxide film. *Adv. Mater.* **2014**, *26*, 3899–3904. [CrossRef] [PubMed]
54. Xue, T.Y.; Cui, X.Q.; Guan, W.M.; Wang, Q.Y.; Liu, C.; Wang, H.T.; Qi, K.; Singh, D.J.; Zheng, W.T. Surface plasmon resonance technique for directly probing the interaction of DNA and graphene oxide and ultra-sensitive biosensing. *Biosens. Bioelectron.* **2014**, *58*, 374–379. [CrossRef] [PubMed]
55. Kundu, B.; Eltohamy, M.; Yadavalli, V.K.; Kundu, S.C.; Kim, H.W. Biomimetic designing of functional silk nanotopography using self assembly. *ACS Appl. Mater. Interfaces* **2016**, *8*, 28458–28467. [CrossRef] [PubMed]
56. Miao, L.; Fan, Q.S.; Zhao, L.L.; Qiao, Q.L.; Zhang, X.Y.; Hou, C.X.; Xu, J.Y.; Luo, Q.; Liu, J.Q. The construction of functional protein nanotubes by small molecule-induced self-assembly of cricoid proteins. *Chem. Commun.* **2016**, *52*, 4092–4095. [CrossRef] [PubMed]
57. Sun, F.D.; Chen, L.; Ding, X.F.; Xu, L.D.; Zhou, X.R.; Wei, P.; Liang, J.F.; Luo, S.Z. High-resolution insights into the stepwise self-assembly of nanofiber from bioactive peptides. *J. Phys. Chem. B* **2017**, *121*, 7421–7430. [CrossRef]
58. Wang, J.H.; Zhao, X.J.; Li, J.F.; Kuang, X.; Fan, Y.Q.; Wei, G.; Su, Z.Q. Electrostatic assembly of peptide nanofiber-biomimetic silver nanowires onto graphene for electrochemical sensors. *ACS Macro Lett.* **2014**, *3*, 529–533. [CrossRef]
59. Palchoudhury, S.; Zhou, Z.Y.; Ramasamy, K.; Okirie, F.; Prevelige, P.E.; Gupta, A. Self-assembly of p22 protein cages with polyamidoamine dendrimer and inorganic nanoparticles. *J. Mater. Res.* **2017**, *32*, 465–472. [CrossRef]
60. Sun, H.C.; Zhang, X.Y.; Miao, L.; Zhao, L.L.; Luo, Q.; Xu, J.Y.; Liu, J.Q. Micelle-induced self-assembling protein nanowires: Versatile supramolecular scaffolds for designing the light-harvesting system. *ACS Nano* **2016**, *10*, 421–428. [CrossRef]
61. Zhao, L.L.; Zou, H.Y.; Zhang, H.; Sun, H.C.; Wang, T.T.; Pan, T.Z.; Li, X.M.; Bai, Y.S.; Ojao, S.P.; Luo, Q.; et al. Enzyme-triggered defined protein nanoarrays: Efficient light-harvesting systems to mimic chloroplasts. *ACS Nano* **2017**, *11*, 938–945. [CrossRef] [PubMed]
62. Liu, L.H.; Li, Z.Y.; Rong, L.; Qin, S.Y.; Lei, Q.; Cheng, H.; Zhou, X.; Zhuo, R.X.; Zhang, X.Z. Self-assembly of hybridized peptide nucleic acid amphiphiles. *ACS Macro Lett.* **2014**, *3*, 467–471. [CrossRef]
63. McGuinness, K.; Nanda, V. Collagen mimetic peptide discs promote assembly of a broad range of natural protein fibers through hydrophobic interactions. *Org. Biomol. Chem.* **2017**, *15*, 5893–5898. [CrossRef] [PubMed]
64. Tanaka, M.; Abiko, S.; Himeiwa, T.; Kinoshita, T. Two-dimensional self-assembly of amphiphilic peptides; adsorption-induced secondary structural transition on hydrophilic substrate. *J. Colloid Interface Sci.* **2015**, *442*, 82–88. [CrossRef] [PubMed]
65. Liao, H.S.; Lin, J.; Liu, Y.; Huang, P.; Jin, A.; Chen, X.Y. Self-assembly mechanisms of nanofibers from peptide amphiphiles in solution and on substrate surfaces. *Nanoscale* **2016**, *8*, 14814–14820. [CrossRef]
66. Yang, H.; Fung, S.Y.; Pritzker, M.; Chen, P. Modification of hydrophilic and hydrophobic surfaces using an ionic-complementary peptide. *PLOS One* **2007**, *2*, e1325. [CrossRef] [PubMed]
67. Anand, B.G.; Dubey, K.; Shekhawat, D.S.; Prajapati, K.P.; Kar, K. Strategically designed antifibrotic gold nanoparticles to prevent collagen fibril formation. *Langmuir* **2017**, *33*, 13252–13261. [CrossRef]
68. Wang, J.Q.; Tao, K.; Yang, Y.Z.; Zhang, L.Y.; Wang, D.; Cao, M.W.; Sun, Y.W.; Xia, D.H. Short peptide mediated self-assembly of platinum nanocrystals with selective spreading property. *RSC Adv.* **2016**, *6*, 58099–58105. [CrossRef]
69. Li, Q.; Liu, L.; Zhang, S.; Xu, M.; Wang, X.Q.; Wang, C.; Besenbacher, F.; Dong, M.D. Modulating a beta(33-42) peptide assembly by graphene oxide. *Chem. Eur. J.* **2014**, *20*, 7236–7240. [CrossRef]
70. Wang, E.; Desai, M.S.; Lee, S.W. Light-controlled graphene-elastin composite hydrogel actuators. *Nano Lett.* **2013**, *13*, 2826–2830. [CrossRef]
71. 71 Lu, C.H.; Li, J.; Zhang, X.L.; Zheng, A.X.; Yang, H.H.; Chen, X.; Chen, G.N. General approach for monitoring peptide-protein interactions based on graphene-peptide complex. *Anal. Chem.* **2011**, *83*, 7276–7282. [CrossRef] [PubMed]

72. Ma, C.; Liu, X.M.; Wu, G.Y.; Zhou, P.; Zhou, Y.T.; Wang, L.; Huang, X. Efficient way to generate protein-based nanoparticles by in-situ photoinitiated polymerization-induced self-assembly. *ACS Macro Lett.* **2017**, *6*, 689–694. [CrossRef]
73. Zhang, Y.; Zhang, J.Y.; Huang, X.L.; Zhou, X.J.; Wu, H.X.; Guo, S.W. Assembly of graphene oxide-enzyme conjugates through hydrophobic interaction. *Small* **2012**, *8*, 154–159. [CrossRef] [PubMed]
74. Law, A.S.Y.; Yeung, M.C.L.; Yam, V.W.W. Arginine-rich peptide-induced supramolecular self-assembly of water-soluble anionic alkynylplatinum(ii) complexes: A continuous and label-free luminescence assay for trypsin and inhibitor screening. *ACS Appl. Mater. Interfaces* **2017**, *9*, 41143–41150. [CrossRef] [PubMed]
75. Wang, M.K.; Lin, Z.H.; Liu, Q.; Jiang, S.; Liu, H.; Su, X.G. DNA-hosted copper nanoclusters/graphene oxide based fluorescent biosensor for protein kinase activity detection. *Anal. Chim. Acta* **2018**, *1012*, 66–73. [CrossRef] [PubMed]
76. Su, Z.Q.; Shen, H.Y.; Wang, H.X.; Wang, J.H.; Li, J.F.; Nienhaus, G.U.; Shang, L.; Wei, G. Motif-designed peptide nanofibers decorated with graphene quantum dots for simultaneous targeting and imaging of tumor cells. *Adv. Funct. Mater.* **2015**, *25*, 5472–5478. [CrossRef]
77. Li, Y.; Zhang, W.S.; Zhang, L.; Li, J.F.; Su, Z.Q.; Wei, G. Sequence-designed peptide nanofibers bridged conjugation of graphene quantum dots with graphene oxide for high performance electrochemical hydrogen peroxide biosensor. *Adv. Mater. Interfaces* **2017**, *4*, 1600895. [CrossRef]
78. Li, D.P.; Zhang, W.S.; Yu, X.Q.; Wang, Z.P.; Su, Z.Q.; Wei, G. When biomolecules meet graphene: From molecular level interactions to material design and applications. *Nanoscale* **2016**, *8*, 19491–19509. [CrossRef]
79. Yun, W.; Xiong, W.; Wu, H.; Fu, M.; Huang, Y.; Liu, X.Y.; Yang, L.Z. Graphene oxide-based fluorescent "turn-on" strategy for hg2+ detection by using catalytic hairpin assembly for amplification. *Sens. Actuat. B* **2017**, *249*, 493–498. [CrossRef]
80. Wang, J.M.; Zhu, H.H.; Xu, Y.H.; Yang, W.R.; Liu, A.; Shan, F.K.; Cao, M.M.; Liu, J.Q. Graphene nanodots encaged 3-d gold substrate as enzyme loading platform for the fabrication of high performance biosensors. *Sens. Actuat. B* **2015**, *220*, 1186–1195. [CrossRef]
81. Huang, J.Y.; Zhao, L.; Lei, W.; Wen, W.; Wang, Y.J.; Bao, T.; Xiong, H.Y.; Zhang, X.H.; Wang, S.F. A high-sensitivity electrochemical aptasensor of carcinoembryonic antigen based on graphene quantum dots-ionic liquid-nafion nanomatrix and dnazyme-assisted signal amplification strategy. *Biosens. Bioelectron.* **2018**, *99*, 28–33. [CrossRef] [PubMed]
82. Li, M.Z.; Pan, Y.X.; Guo, X.Y.; Liang, Y.H.; Wu, Y.P.; Wen, Y.; Yang, H.F. Pt/single-stranded DNA/graphene nanocomposite with improved catalytic activity and co tolerance. *J. Mater. Chem. A* **2015**, *3*, 10353–10359. [CrossRef]
83. Shen, W.L.; Liu, Q.; Ding, B.Q.; Zhu, C.Q.; Shen, Z.Y.; Seeman, N.C. Facilitation of DNA self-assembly by relieving the torsional strains between building blocks. *Org. Biomol. Chem.* **2017**, *15*, 465–469. [CrossRef] [PubMed]
84. Zhang, C.; Yang, J.; Jiang, S.X.; Liu, Y.; Yan, H. Dnazyme-based logic gate-mediated DNA self-assembly. *Nano Lett.* **2016**, *16*, 736–741. [CrossRef]
85. Elbaz, J.; Yin, P.; Voigt, C.A. Genetic encoding of DNA nanostructures and their self-assembly in living bacteria. *Nat. Commun.* **2016**, *7*, 11179. [CrossRef]
86. Zhang, F.; Yan, H. DNA self-assembly scaled up. *Nature* **2017**, *552*, 185. [CrossRef]
87. Li, Z.; Liu, M.H.; Wang, L.; Nangreave, J.; Yan, H.; Liu, Y. Molecular behavior of DNA origami in higher-order self-assembly. *J. Am. Chem. Soc.* **2010**, *132*, 13545–13552. [CrossRef] [PubMed]
88. Idili, A.; Vallee-Belisle, A.; Ricci, F. Programmable pH-triggered DNA nanoswitches. *J. Am. Chem. Soc.* **2014**, *136*, 5836–5839. [CrossRef]
89. Kuzyk, A.; Laitinen, K.T.; Torma, P. DNA origami as a nanoscale template for protein assembly. *Nanotechnology* **2009**, *20*, 235305. [CrossRef] [PubMed]
90. Pal, S.; Deng, Z.T.; Ding, B.Q.; Yan, H.; Liu, Y. DNA-origami-directed self-assembly of discrete silver-nanoparticle architectures. *Angew. Chem. Int. Ed.* **2010**, *49*, 2700–2704. [CrossRef]
91. Li, H.; Zhang, K.M.; Pi, F.M.; Guo, S.J.; Shlyakhtenko, L.; Chiu, W.; Shu, D.; Guo, P.X. Controllable self-assembly of rna tetrahedrons with precise shape and size for cancer targeting. *Adv. Mater.* **2016**, *28*, 7501–7507. [CrossRef] [PubMed]
92. Boerneke, M.A.; Dibrov, S.M.; Hermann, T. Crystal-structure-guided design of self-assembling RNA nanotriangles. *Angew. Chem. Int. Ed.* **2016**, *55*, 4097–4100. [CrossRef] [PubMed]

93. Stewart, J.M.; Subramanian, H.K.K.; Franco, E. Self-assembly of multi-stranded rna motifs into lattices and tubular structures (vol 19, pg 5449, 2017). *Nucleic Acids Res.* **2017**, *45*, 5628. [CrossRef] [PubMed]
94. Berger, O.; Adler-Abramovich, L.; Levy-Sakin, M.; Grunwald, A.; Liebes-Peer, Y.; Bachar, M.; Buzhansky, L.; Mossou, E.; Forsyth, V.T.; Schwartz, T.; et al. Light-emitting self-assembled peptide nucleic acids exhibit both stacking interactions and watson-crick base pairing. *Nat. Nanotechnol.* **2015**, *10*, 353–360. [CrossRef] [PubMed]
95. Liljestrom, V.; Mikkila, J.; Kostiainen, M.A. Self-assembly and modular functionalization of three-dimensional crystals from oppositely charged proteins. *Nat. Commun.* **2014**, *5*, 4445. [CrossRef] [PubMed]
96. Haburcak, R.; Shi, J.F.; Du, X.W.; Yuan, D.; Xu, B. Ligand-receptor interaction modulates the energy landscape of enzyme-instructed self-assembly of small molecules. *J. Am. Chem. Soc.* **2016**, *138*, 15397–15404. [CrossRef] [PubMed]
97. Bachmann, S.J.; Petitzon, M.; Mognetti, B.M. Bond formation kinetics affects self-assembly directed by ligand-receptor interactions. *Soft Matter* **2016**, *12*, 9585–9592. [CrossRef] [PubMed]
98. Wang, S.P.; Mamedova, N.; Kotov, N.A.; Chen, W.; Studer, J. Antigen/antibody immunocomplex from cdte nanoparticle bioconjugates. *Nano Lett.* **2002**, *2*, 817–822. [CrossRef]
99. Kominami, H.; Kobayashi, K.; Ido, S.; Kimiya, H.; Yamada, H. Immunoactivity of self-assembled antibodies investigated by atomic force microscopy. *RSC Adv.* **2018**, *8*, 29378–29384. [CrossRef]
100. Fong, W.K.; Negrini, R.; Vallooran, J.J.; Mezzenga, R.; Boyd, B.J. Responsive self-assembled nanostructured lipid systems for drug delivery and diagnostics. *J. Colloid Interface Sci.* **2016**, *484*, 320–339. [CrossRef]
101. Zhou, Y.X.; Briand, V.A.; Sharma, N.; Ahn, S.K.; Kasi, R.M. Polymers comprising cholesterol: Synthesis, self-assembly, and applications. *Materials* **2009**, *2*, 636–660. [CrossRef]
102. Conn, C.E.; Drummond, C.J. Nanostructured bicontinuous cubic lipid self-assembly materials as matrices for protein encapsulation. *Soft Matter* **2013**, *9*, 3449–3464. [CrossRef]
103. Zerkoune, L.; Lesieur, S.; Putaux, J.L.; Choisnard, L.; Geze, A.; Wouessidjewe, D.; Angelov, B.; Vebert-Nardin, C.; Doutch, J.; Angelova, A. Mesoporous self-assembled nanoparticles of biotransesterified cyclodextrins and nonlamellar lipids as carriers of water-insoluble substances. *Soft Matter* **2016**, *12*, 7539–7550. [CrossRef] [PubMed]
104. Yang, H.K.; Ren, L.J.; Wu, H.; Wang, W. Self-assembly of the polyoxometalate-cholesterol conjugate into microrods or nanoribbons regulated by thermodynamics. *New J. Chem.* **2016**, *40*, 954–961. [CrossRef]
105. Engberg, K.; Waters, D.J.; Kelmanovich, S.; Parke-Houben, R.; Hartmann, L.; Toney, M.F.; Frank, C.W. Self-assembly of cholesterol tethered within hydrogel networks. *Polymer* **2016**, *84*, 371–382. [CrossRef]
106. Lei, H.R.; Liu, J.; Yan, J.L.; Quan, J.M.; Fang, Y. Luminescent helical nanofiber self-assembled from a cholesterol-based metalloamphiphile and its application in DNA conformation recognition. *Langmuir* **2016**, *32*, 10350–10357. [CrossRef] [PubMed]
107. Silva, N.H.C.S.; Pinto, R.J.B.; Freire, C.S.R.; Marrucho, I.M. Production of lysozyme nanofibers using deep eutectic solvent aqueous solutions. *Colloids Surf. B* **2016**, *147*, 36–44. [CrossRef] [PubMed]
108. Dinesh, B.; Squillaci, M.A.; Menard-Moyon, C.; Samori, P.; Bianco, A. Self-assembly of diphenylalanine backbone homologues and their combination with functionalized carbon nanotubes. *Nanoscale* **2015**, *7*, 15873–15879. [CrossRef] [PubMed]
109. Moyer, T.J.; Finbloom, J.A.; Chen, F.; Toft, D.J.; Cryns, V.L.; Stupp, S.I. pH and amphiphilic structure direct supramolecular behavior in biofunctional assemblies. *J. Am. Chem. Soc.* **2014**, *136*, 14746–14752. [CrossRef] [PubMed]
110. Cote, Y.; Fu, I.W.; Dobson, E.T.; Goldberger, J.E.; Nguyen, H.D.; Shen, J.K. Mechanism of the pH-controlled self-assembly of nanofibers from peptide amphiphiles. *J. Phys. Chem. C* **2014**, *118*, 16272–16278. [CrossRef]
111. Jana, P.; Ehlers, M.; Zellermann, E.; Samanta, K.; Schmuck, C. Ph-controlled formation of a stable beta-sheet and amyloid-like fibers from an amphiphilic peptide: The importance of a tailor-made binding motif for secondary structure formation. *Angew. Chem. Int. Ed.* **2016**, *55*, 15287–15291. [CrossRef] [PubMed]
112. Hsieh, M.C.; Liang, C.; Mehta, A.K.; Lynn, D.G.; Grover, M.A. Multistep conformation selection in amyloid assembly. *J. Am. Chem. Soc.* **2017**, *139*, 17007–17010. [CrossRef] [PubMed]
113. Ghosh, A.; Haverick, M.; Stump, K.; Yang, X.Y.; Tweedle, M.F.; Goldberger, J.E. Fine-tuning the pH trigger of self-assembly. *J. Am. Chem. Soc.* **2012**, *134*, 3647–3650. [CrossRef] [PubMed]
114. Chen, Y.R.; Gan, H.X.; Tong, Y.W. pH-controlled hierarchical self-assembly of peptide amphiphile. *Macromolecules* **2015**, *48*, 2647–2653. [CrossRef]

115. Larnaudie, S.C.; Brendel, J.C.; Jolliffe, K.A.; Perrier, S. pH-responsive, amphiphilic core-shell supramolecular polymer brushes from cyclic peptide-polymer conjugates. *ACS Macro Lett.* **2017**, *6*, 1347–1351. [CrossRef]
116. Reinecke, A.; Brezesinski, G.; Harrington, M.J. pH-responsive self-organization of metal-binding protein motifs from biomolecular junctions in mussel byssus. *Adv. Mater. Interfaces* **2017**, *4*, 1600416. [CrossRef]
117. Brodin, J.D.; Ambroggio, X.I.; Tang, C.Y.; Parent, K.N.; Baker, T.S.; Tezcan, F.A. Metal-directed, chemically tunable assembly of one-, two- and three-dimensional crystalline protein arrays. *Nat. Chem.* **2012**, *4*, 375–382. [CrossRef] [PubMed]
118. Brodin, J.D.; Carr, J.R.; Sontz, P.A.; Tezcan, F.A. Exceptionally stable, redox-active supramolecular protein assemblies with emergent properties. *Proc. Natl. Acad. Sci. USA* **2014**, *111*, 2897–2902. [CrossRef] [PubMed]
119. Elbaz, J.; Wang, Z.G.; Orbach, R.; Willner, I. pH-stimulated concurrent mechanical activation of two DNA "tweezers". A "set-reset" logic gate system. *Nano Lett.* **2009**, *9*, 4510–4514. [CrossRef] [PubMed]
120. Lu, C.H.; Cecconello, A.; Elbaz, J.; Credi, A.; Willner, I. A three-station DNA catenane rotary motor with controlled directionality. *Nano Lett.* **2013**, *13*, 2303–2308. [CrossRef] [PubMed]
121. Wu, N.; Willner, I. pH-stimulated reconfiguration and structural isomerization of origami dimer and trimer systems. *Nano Lett.* **2016**, *16*, 6650–6655. [CrossRef] [PubMed]
122. Ozkan, A.D.; Tekinay, A.B.; Guler, M.O.; Tekin, E.D. Effects of temperature, pH and counterions on the stability of peptide amphiphile nanofiber structures. *RSC Adv.* **2016**, *6*, 104201–104214. [CrossRef]
123. Yu, M.; Tang, T.; Takasu, A.; Higuchi, M. pH- and thermo-induced morphological changes of an amphiphilic peptide-grafted copolymer in solution. *Polym. J.* **2014**, *46*, 52–58. [CrossRef]
124. Castelletto, V.; Cheng, G.; Stain, C.; Connon, C.J.; Hamley, I.W. Self-assembly of a peptide amphiphile containing l-carnosine and its mixtures with a multilamellar vesicle forming lipid. *Langmuir* **2012**, *28*, 11599–11608. [CrossRef] [PubMed]
125. Hamley, I.W.; Dehsorkhi, A.; Castelletto, V.; Furzeland, S.; Atkins, D.; Seitsonen, J.; Ruokolainen, J. Reversible helical unwinding transition of a self-assembling peptide amphiphile. *Soft Matter* **2013**, *9*, 9290–9293. [CrossRef]
126. Jiang, L.D.; Bonde, J.S.; Ye, L. Temperature and pH controlled self-assembly of a protein-polymer biohybrid. *Macromol. Chem. Phys.* **2018**, *219*, 1700597.
127. Huang, Y.J.; Mai, Y.Y.; Yang, X.W.; Beser, U.; Liu, J.Z.; Zhang, F.; Yan, D.Y.; Mullen, K.; Feng, X.L. Temperature-dependent multidimensional self-assembly of polyphenylene-based "rod-coil" graft polymers. *J. Am. Chem. Soc.* **2015**, *137*, 11602–11605. [CrossRef]
128. Zhong, J.; Liu, X.W.; Wei, D.X.; Yan, J.; Wang, P.; Sun, G.; He, D.N. Effect of incubation temperature on the self-assembly of regenerated silk fibroin: A study using afm. *Int. J. Biol. Macromol.* **2015**, *76*, 195–202. [CrossRef]
129. Putri, R.M.; Cornelissen, J.J.L.M.; Koay, M.S.T. Self-assembled cage-like protein structures. *ChemPhysChem* **2015**, *16*, 911–918. [CrossRef]
130. Semerdzhiev, S.A.; Dekker, D.R.; Subramaniam, V.; Claessens, M.M.A.E. Self-assembly of protein fibrils into suprafibrillar aggregates: Bridging the nano- and mesoscale. *ACS Nano* **2014**, *8*, 5543–5551. [CrossRef]
131. Dai, B.; Li, D.; Xi, W.; Luo, F.; Zhang, X.; Zou, M.; Cao, M.; Hu, J.; Wang, W.Y.; Wei, G.H.; et al. Tunable assembly of amyloid-forming peptides into nanosheets as a retrovirus carrier. *Proc. Natl. Acad. Sci. USA* **2015**, *112*, 2996–3001. [CrossRef] [PubMed]
132. Liu, L.F.; Li, Y.L.; Wang, Y.; Zheng, J.W.; Mao, C.D. Regulating DNA self-assembly by DNA-surface interactions. *ChemBioChem* **2017**, *18*, 2404–2407. [CrossRef] [PubMed]
133. Garmann, R.F.; Comas-Garcia, M.; Gopal, A.; Knobler, C.M.; Gelbart, W.M. The assembly pathway of an icosahedral single-stranded rna virus depends on the strength of inter-subunit attractions. *J. Mol. Biol.* **2014**, *426*, 1050–1060. [CrossRef] [PubMed]
134. Yang, S.; Liu, W.Y.; Nixon, R.; Wang, R.S. Metal-ion responsive reversible assembly of DNA origami dimers: G-quadruplex induced intermolecular interaction. *Nanoscale* **2018**, *10*, 3626–3630. [CrossRef] [PubMed]
135. Yan, X.H.; Cui, Y.; He, Q.; Wang, K.W.; Li, J.B. Organogels based on self-assembly of diphenylalanine peptide and their application to immobilize quantum dots. *Chem. Mater.* **2008**, *20*, 1522–1526. [CrossRef]
136. Zhu, P.L.; Yan, X.H.; Su, Y.; Yang, Y.; Li, J.B. Solvent-induced structural transition of self-assembled dipeptide: From organogels to microcrystals. *Chem. Eur. J.* **2010**, *16*, 3176–3183. [CrossRef] [PubMed]
137. Huang, R.L.; Qi, W.; Su, R.X.; Zhao, J.; He, Z.M. Solvent and surface controlled self-assembly of diphenylalanine peptide: From microtubes to nanofibers. *Soft Matter* **2011**, *7*, 6418–6421. [CrossRef]

138. Mason, T.O.; Chirgadze, D.Y.; Levin, A.; Adler-Abramovich, L.; Gazit, E.; Knowles, T.P.J.; Buell, A.K. Expanding the solvent chemical space for self-assembly of dipeptide nanostructures. *ACS Nano* **2014**, *8*, 1243–1253. [CrossRef] [PubMed]
139. Su, Y.; Yan, X.H.; Wang, A.H.; Fei, J.B.; Cui, Y.; He, Q.; Li, J.B. A peony-flower-like hierarchical mesocrystal formed by diphenylalanine. *J. Mater. Chem.* **2010**, *20*, 6734–6740. [CrossRef]
140. Ryu, J.; Park, C.B. High-temperature self-assembly of peptides into vertically well-aligned nanowires by aniline vapor. *Adv. Mater.* **2008**, *20*, 3754–3758. [CrossRef]
141. Helbing, C.; Deckert-Gaudig, T.; Firkowska-Boden, I.; Wei, G.; Deckert, V.; Jandt, K.D. Protein handshake on the nanoscale: How albumin and hemoglobin self-assemble into nanohybrid fibers. *ACS Nano* **2018**, *12*, 1211–1219. [CrossRef] [PubMed]
142. Wang, J.; Liu, K.; Yan, L.Y.; Wang, A.H.; Bai, S.; Yan, X.H. Trace solvent as a predominant factor to tune dipeptide self-assembly. *ACS Nano* **2016**, *10*, 2138–2143. [CrossRef] [PubMed]
143. Fu, I.W.; Markegard, C.B.; Nguyen, H.D. Solvent effects on kinetic mechanisms of self-assembly by peptide amphiphiles via molecular dynamics simulations. *Langmuir* **2015**, *31*, 315–324. [CrossRef] [PubMed]
144. He, H.J.; Xu, B. Instructed-assembly (IA): A molecular process for controlling cell fate. *Bull. Chem. Soc. Jpn.* **2018**, *91*, 900–906. [CrossRef] [PubMed]
145. Hahn, M.E.; Gianneschi, N.C. Enzyme-directed assembly and manipulation of organic nanomaterials. *Chem. Commun.* **2011**, *47*, 11814–11821. [CrossRef] [PubMed]
146. Yang, Z.M.; Gu, H.W.; Fu, D.G.; Gao, P.; Lam, J.K.; Xu, B. Enzymatic formation of supramolecular hydrogels. *Adv. Mater.* **2004**, *16*, 1440–1444. [CrossRef]
147. Amir, R.J.; Zhong, S.; Pochan, D.J.; Hawker, C.J. Enzymatically triggered self-assembly of block copolymers. *J. Am. Chem. Soc.* **2009**, *131*, 13949–13951. [CrossRef]
148. Guilbaud, J.B.; Vey, E.; Boothroyd, S.; Smith, A.M.; Ulijn, R.V.; Saiani, A.; Miller, A.F. Enzymatic catalyzed synthesis and triggered gelation of ionic peptides. *Langmuir* **2010**, *26*, 11297–11303. [CrossRef] [PubMed]
149. Xu, J.X.; Zhou, Z.; Wu, B.; He, B.F. Enzymatic formation of a novel cell-adhesive hydrogel based on small peptides with a laterally grafted l-3,4-dihydroxyphenylalanine group. *Nanoscale* **2014**, *6*, 1277–1280. [CrossRef] [PubMed]
150. Chien, M.P.; Rush, A.M.; Thompson, M.P.; Gianneschi, N.C. Programmable shape-shifting micelles. *Angew. Chem. Int. Ed.* **2010**, *49*, 5076–5080. [CrossRef] [PubMed]
151. Heck, T.; Faccio, G.; Richter, M.; Thony-Meyer, L. Enzyme-catalyzed protein crosslinking. *Appl. Microbiol. Biotechnol.* **2013**, *97*, 461–475. [CrossRef]
152. Yuan, D.; Shi, J.F.; Du, X.W.; Huang, Y.B.; Gao, Y.; Baoum, A.A.; Xu, B. The enzyme-instructed assembly of the core of yeast prion sup35 to form supramolecular hydrogels. *J. Mater. Chem. B* **2016**, *4*, 1318–1323. [CrossRef] [PubMed]
153. He, H.J.; Wang, H.M.; Zhou, N.; Yang, D.S.; Xu, B. Branched peptides for enzymatic supramolecular hydrogelation. *Chem. Commun.* **2018**, *54*, 86–89. [CrossRef]
154. He, H.J.; Wang, J.Q.; Wang, H.M.; Zhou, N.; Yang, D.; Green, D.R.; Xu, B. Enzymatic cleavage of branched peptides for targeting mitochondria. *J. Am. Chem. Soc.* **2018**, *140*, 1215–1218. [CrossRef] [PubMed]
155. Zhou, J.; Du, X.W.; Chen, X.Y.; Wang, J.Q.; Zhou, N.; Wu, D.F.; Xu, B. Enzymatic self-assembly confers exceptionally strong synergism with nf-kappa b targeting for selective necroptosis of cancer cells. *J. Am. Chem. Soc.* **2018**, *140*, 2301–2308. [CrossRef] [PubMed]
156. Qi, J.L.; Yan, Y.F.; Cheng, B.C.; Deng, L.F.; Shao, Z.W.; Sun, Z.L.; Li, X.M. Enzymatic formation of an injectable hydrogel from a glycopeptide as a biomimetic scaffold for vascularization. *ACS Appl. Mater. Interfaces* **2018**, *10*, 6180–6189. [CrossRef] [PubMed]
157. Muraoka, T.; Cui, H.; Stupp, S.I. Quadruple helix formation of a photoresponsive peptide amphiphile and its light-triggered dissociation into single fibers. *J. Am. Chem. Soc.* **2008**, *130*, 2946–2947. [CrossRef]
158. Ma, H.C.; Fei, J.B.; Li, Q.; Li, J.B. Photo-induced reversible structural transition of cationic diphenylalanine peptide self-assembly. *Small* **2015**, *11*, 1787–1791. [CrossRef]
159. Tanaka, F.; Mochizuki, T.; Liang, X.G.; Asanuma, H.; Tanaka, S.; Suzuki, K.; Kitamura, S.; Nishikawa, A.; Ui-Tei, K.; Hagiya, M. Robust and photocontrollable DNA capsules using azobenzenes. *Nano Lett.* **2010**, *10*, 3560–3565. [CrossRef] [PubMed]
160. Yang, Y.Y.; Endo, M.; Hidaka, K.; Sugiyama, H. Photo-controllable DNA origami nanostructures assembling into predesigned multiorientational patterns. *J. Am. Chem. Soc.* **2012**, *134*, 20645–20653. [CrossRef] [PubMed]

161. Suzuki, Y.; Endo, M.; Yang, Y.Y.; Sugiyama, H. Dynamic assembly/disassembly processes of photoresponsive DNA origami nanostructures directly visualized on a lipid membrane surface. *J. Am. Chem. Soc.* **2014**, *136*, 1714–1717. [CrossRef] [PubMed]
162. Sun, Y.Q.; Zhang, Y.N.; Tian, L.L.; Zhao, Y.Y.; Wu, D.N.; Xue, W.; Ramakrishna, S.; Wu, W.T.; He, L.M. Self-assembly behaviors of molecular designer functional rada16-i peptides: Influence of motifs, pH, and assembly time. *Biomed. Mater.* **2017**, *12*, 015007. [CrossRef] [PubMed]
163. Shao, Y.; Jia, H.Y.; Cao, T.Y.; Liu, D.S. Supramolecular hydrogels based on DNA self-assembly. *Acc. Chem. Res.* **2017**, *50*, 659–668. [CrossRef] [PubMed]
164. Li, K.; Zhang, Z.F.; Li, D.P.; Zhang, W.S.; Yu, X.Q.; Liu, W.; Gong, C.C.; Wei, G.; Su, Z.Q. Biomimetic Ultralight, Highly Porous, Shape-Adjustable, and Biocompatible 3D Graphene Minerals via Incorporation of Self-Assembled Peptide Nanosheets. *Adv. Funct. Mater.* **2018**, *28*, 1801056. [CrossRef]
165. Hamley, I.W. Small bioactive peptides for biomaterials design and therapeutics. *Chem. Rev.* **2017**, *117*, 14015–14041. [CrossRef]
166. Gong, C.C.; Sun, S.W.; Zhang, Y.J.; Sun, L.; Su, Z.Q.; Wu, A.G.; Wei, G. Hierarchical nanomaterials via biomolecular self-assembly and bioinspiration for energy and environmental applications. *Nanocale* **2019**, *11*. [CrossRef]

nanomaterials

MDPI

Article

Synthesis, Self-Assembly, and Drug-Release Properties of New Amphipathic Liquid Crystal Polycarbonates

Yujiao Xie [1], Xiaofeng Liu [1], Zhuang Hu [1], Zhipeng Hou [1], Zhihao Guo [1], Zhangpei Chen [1], Jianshe Hu [1,*] and Liqun Yang [2,*]

[1] Center for Molecular Science and Engineering, College of Science, Northeastern University, Shenyang 110819, China; xieyujiao5819573@gmail.com (Y.X.); 1510048@stu.neu.edu.cn (X.L.); 1710059@stu.neu.edu.cn (Z.H.); 1670175@stu.neu.edu.cn (Z.H.); 1610046@stu.neu.edu.cn (Z.G.); chenzhangpei@mail.neu.edu.cn (Z.C.)

[2] Key Laboratory of Reproductive Health, Liaoning Research Institute of Family Planning, Shenyang 110031, China

* Correspondence: hujs@mail.neu.edu.cn (J.H.); yanglq@lnszjk.com.cn (L.Y.)

Received: 28 January 2018; Accepted: 25 March 2018; Published: 27 March 2018

Abstract: New amphiphilic liquid crystal (LC) polycarbonate block copolymers containing side-chain cholesteryl units were synthesized. Their structure, thermal stability, and LC phase behavior were characterized with Fourier transform infrared (FT-IR) spectrum, ^{1}H NMR, gel permeation chromatographic (GPC), thermogravimetric analysis (TGA), differential scanning calorimetry (DSC), polarizing optical microscope (POM), and XRD methods. The results demonstrated that the LC copolymers showed a double molecular arrangement of a smectic A phase at room temperature. With the elevating of LC unit content in such LC copolymers, the corresponding properties including decomposition temperature (T_d), glass temperature (T_g), and isotropic temperature (T_i) increased. The LC copolymers showed pH-responsive self-assembly behavior under the weakly acidic condition, and with more side-chain LC units, the self-assembly process was faster, and the formed particle size was smaller. It indicated that the self-assembly driving force was derived from the orientational ability of LC. The particle size and morphologies of self-assembled microspheres loaded with doxorubicin (DOX), together with drug release tracking, were evaluated by dynamic light scattering (DLS), SEM, and UV–vis spectroscopy. The results showed that DOX could be quickly released in a weakly acidic environment due to the pH response of the self-assembled microspheres. This would offer a new strategy for drug delivery in clinic applications.

Keywords: amphipathic polycarbonates; cholesteryl; liquid crystal; self-assembly; drug release

1. Introduction

Biodegradable aliphatic polyesters have received more and more attention in recent years due to their good biodegradability and biocompatibility [1,2]. Among them, four class of polyesters, namely polyglycolide (PGA), poly(1,3-trimethylene carbonate) (PTMC), polylactide (PLA), poly(caprolactone) (PCL), and their copolymers, are the most widely studied and utilized in biomedical fields. However, exploration of their applications has been hampered by several limitations owing to their specific properties, for example, the PGA degradation rate is too fast with respect to the required application time of materials [3–5]; and PLA can produce acid degradation products which may cause aseptic inflammation [6]. Although there is no acid degradation product during the degradation of PTMC [7], it lacks functional groups to link bioactive primitives. Therefore, 5-benzyloxy-trimethylene carbonate (BTMC) had been selected as the cyclic monomer to form polycarbonate copolymers containing hydroxyl groups [8–10].

It has been shown that various sensory mechanisms of organisms are related to the self-assembly behavior of liquid crystals (LCs) in vivo [11]. Another important point is that LCs can self-assemble to form a variety of ordered structures which are responsive to temperature, pH, stress, magnetic field, and other external conditions [12–14]; as a result, LC structures present considerable advantages in the transport and delivery of diverse active molecules and drugs [15]. Cholesterol is an indispensable material in the animal and human body, and its derivatives were found to have LC properties at earlier time [16–18]. Therefore, as a natural biomesogen, cholesterol has become an attractive candidate for forming new smart, responsive, and biodegradable LC polymer materials. Nowadays, it has been reported that the cholesteryl groups could be employed as mesogenic units to fabricate side-chain [19–23] and main-chain LC polymers [24–28]. However, there is limited research on aliphatic polycarbonates with side-chain cholesteryl LC units [21,29].

The use of polymers for drug delivery is one of the most important drug delivery systems, and the drug-loaded nanoparticle formed with polymers are faster and more convenient in medical treatment since they can be injected to avoid major surgery [4,30,31]. The nanoparticles with tunable properties may be combined by the self-assembly of LC groups and amphiphiles of appropriate chemical structures, such as PEGylated polycarbonates [32,33]. Furthermore, LC polymers, created by the principles of self-assembly and nanocarriers, have been studied with the incorporation of different bioactive macromolecules such as siRNA, plasmid DNA, peptides, and proteins [34–38]. However, there is rarely research on the self-assembly of LC polymers with cholesterol [14]. Thus, it is very significant to the scientific community to study biodegradable aliphatic polycarbonates with cholesteryl LC groups, and their potential medical and clinical application value.

In the previous work, we reported the synthesis and self-assembled morphology of new side-chain diosgenin-functionalized block copolymers with an aliphatic polycarbonate backbone [39]. In this study, a series of new cholesterol-functionalized amphipathic LC copolymers based on aliphatic polycarbonates were synthesized. The design aim is for the hydrophilic segment to be introduced into the aliphatic polycarbonate to improve the hydrophilicity of the copolymer, and the hydroxyl groups are introduced to create reactive sites which can link cholesteryl LC units. The synthesis, thermal stability, and phase behavior of the obtained LC copolymers were characterized by FT-IR, ^{1}H NMR, gel permeation chromatographic (GPC), thermogravimetric analysis (TGA), differential scanning calorimetry (DSC), polarizing optical microscope (POM), and X-ray diffraction (XRD) measurements. Subsequently, the pH-responsive self-assembly process of the target copolymers was studied by UV–vis, and their particle size and morphologies were characterized by dynamic light scattering (DLS) and scanning electron microscope (SEM), respectively. Doxorubicin (DOX), a broad-spectrum anticancer drug, was chosen as the drug model to prepare drug-loaded LC copolymer microspheres via a dialysis method. The particle size and morphologies of DOX-loaded microspheres were detected by DLS and SEM, respectively. The drug-loading efficiency and pH-responsive drug release behavior were monitored by UV–vis. It could serve as a potential biomedical material as well as a drug delivery carrier for tumor therapy.

2. Experimental Method

2.1. Materials

All chemicals were obtained from the indicated sources. 1,3-Trimethylene carbonate (TMC) was purchased from Daigang Biomaterial Co. Ltd. (Jinan, China), and recrystallized twice with acetic ether and then dried 24 h in a vacuum before polymerization. Stannous octanoate ($Sn(Oct)_2$) and methoxypolyethylene glycols ($mPEG_{43}$) were purchased from Aldrich and used without purification. Doxorubicin hydrochloride (DOX·HCl) was purchased from Ark Pharm, Inc. (Arlington Heights, IL, USA) Triethylamine (TEA) was purchased from Sinopharm Chemical Reagent Co., Ltd. (Shenyang, China). Tetrahydrofuran (THF) and toluene were dried by treatment with Na and distilled before use. All other solvents and reagents used were purified by standard methods.

2.2. Measurements

FT-IR spectra. FT-IR spectra were obtained using a PerkinElmer spectrum One (B) spectrometer (PerkinElmer, Foster City, CA, USA). Solid samples were pressed into KBr pellets and liquid samples were drop-casted on KBr pellets.

^{1}H NMR spectra. ^{1}H NMR spectra were obtained using a Bruker ARX 600 (Karlsruhe, Germany) high-resolution NMR spectrometer, and chemical shifts were reported in ppm with tetramethylsilane (TMS) as an internal standard.

Gel permeation chromatographic (GPC). GPC were carried out at room temperature on a Waters 1515 instrument (Shanghai, China) calibrated with THF as an eluent and polystyrene as the standard.

Thermogravimetric analysis (TGA). The thermal decomposition temperature was measured under a nitrogen atmosphere with a Netzsch 209C thermogravimetric analysis (Hanau, Germany) at a heating rate of 20 °C/min.

Differential scanning calorimetry (DSC). The phase transition temperature was determined with a Netzsch 204 (Netzsch, Hanau, Germany) DSC equipped with a cooling system at a heating and cooling rate of 10 °C/min in a nitrogen atmosphere.

Polarizing optical microscope (POM). The optical textures were observed with a Leica DMRX (Leica, Wetzlar, Germany) POM equipped with a Linkam THMSE-600 (Linkam, London, UK) cool and hot stage.

X-ray diffraction (XRD). XRD measurements were performed with nickel-filtered Cu-Kα radiation (λ = 1.54 Å) with a Bruker D8 Advance (Karlsruhe, Germany) powder diffractometer. The temperature-dependent X-ray measurements were carried out in the heating process.

Dynamic light scattering (DLS). The particle size and size distribution of the micelles were determined with DLS using a Zetasizer Nano S instrument (Malvern Instruments Ltd., Worcestershire, UK) with a He–Ne laser (633 nm) set at 173° for the scattering angle at 25 °C.

Scanning electron microscope (SEM). The morphology of the micelles was examined using a Hitachi X650 (Tokyo, Japan). The SEM samples were prepared by depositing several drops of the samples suspension onto the surface of cleaned aluminum foil, and the samples were quenched by liquid nitrogen, then freeze-dried in a vacuum at −50 °C for 24 h. The samples were coated with a thin film of gold before measuring.

UV–vis. Light transmittance test and drug absorbance were measured using a TU-1901 double-beam UV–visible spectrophotometer (Beijing Purkinje General Instrument Co., Ltd., Beijing, China).

2.3. Synthesis of the Block Copolymers

The synthetic route of the block copolymers is outlined in Scheme 1. The cyclic monomer BTMC and the chiral LC monomer 6-cholesteroxy-6-oxocaproic acid (C) were synthesized according to our previous works [6,29].

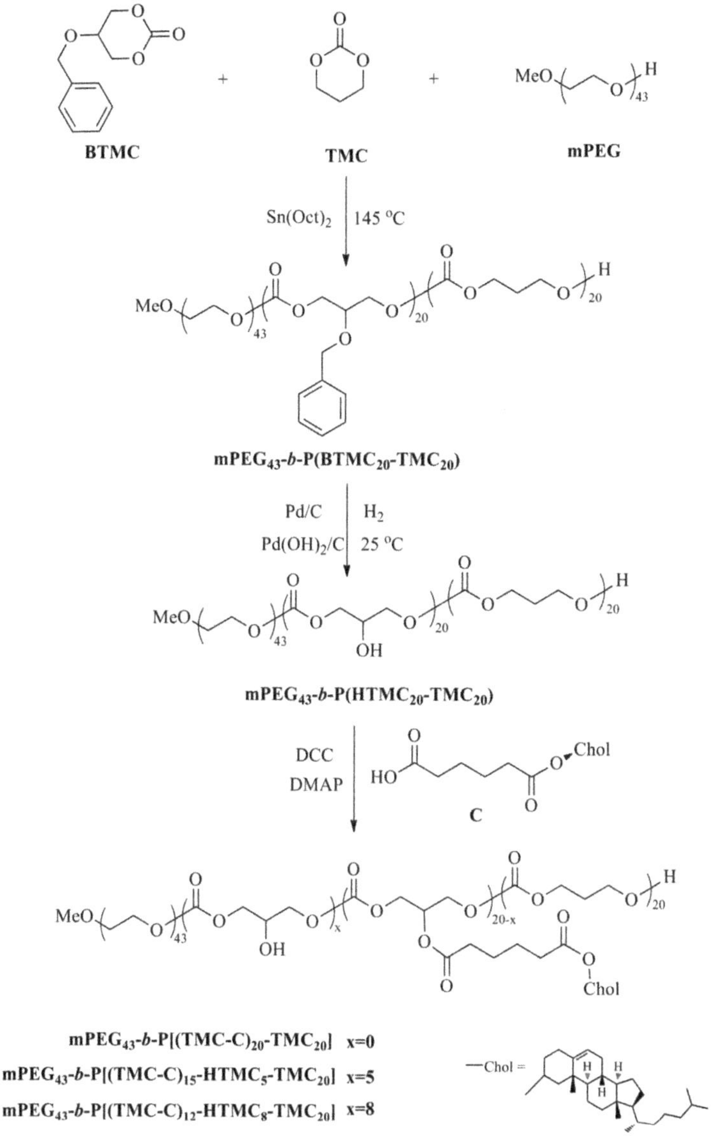

Scheme 1. Synthetic route of the copolymers (C: 6-cholesteroxy-6-oxocaproic acid).

2.3.1. Synthesis of mPEG$_{43}$-*b*-P(BTMC$_{20}$-TMC$_{20}$)

BTMC (10.4 g, 0.05 mol), TMC (4.9 g, 0.05 mol), and mPEG (3.8 g, 0.002 mol) were placed in a polymerization flask, and then $Sn(Oct)_2$ toluene solution (0.13 mol/L, 1.6 mL) as a catalyst was added to the above flask. The flask was sealed in a vacuum and placed in an oil bath kept at 145 °C. The reaction was maintained for 24 h. After the completion of polymerization, the crude product was purified by dissolving it in dichloromethane and precipitated in methanol, and then dried in a vacuum until the sample mass was constant.

IR (KBr, cm^{-1}): 2961, 2875 ($-CH_2-$, $-CH_3$); 1752 (C=O); 1531, 1455 (–Ph); 1241 (C–O–C); ^{1}H NMR (δ, $CDCl_3$, 600 MHz): δ = 7.26 (Ph–*H*), δ = 4.65 (Ph–CH_2O–), δ = 4.25 (–OCH_2–CH–CH_2O), δ = 3.85 (–OCH_2–*CH*–CH_2O– in PBTMC), δ = 3.64 (–OCH_2– in mPEG), δ = 2.03 (–OCH_2–CH_2–OCH_2– in PTMC).

2.3.2. Synthesis of mPEG$_{43}$-*b*-P(HTMC$_{20}$-TMC$_{20}$)

mPEG$_{43}$-*b*-P(BTMC$_{20}$-TMC$_{20}$) (2.1 g) was dissolved in methanol/THF (1:1, 100 mL) and placed in a 250 mL three-necked flask with a magnetic stirrer. Pd/C (0.10 g, 5%) and $Pd(OH)_2$/C (0.10 g, 5%) were added to the above solution. The reaction mixture was stirred for 48 h in a hydrogen system at 25 °C. After the reaction, Pd/C and $Pd(OH)_2$/C were filtered out. The filtrate was evaporated to dryness, and then dried in a vacuum until constant weight was obtained.

IR (KBr, cm^{-1}): 3376 (–OH); 2955, 2853 ($-CH_2-$, $-CH_3$); 1750 (C=O); 1252 (C–O–C); ^{1}H NMR (δ, DMSO, 600 MHz): δ = 5.45 (–OH), δ = 4.09 (–OCH_2–CH–CH_2O–), δ = 3.93 (–OCH_2–*CH*–CH_2O– in PHTMC), δ = 3.51 (–OCH_2– in mPEG), δ = 1.95 (–OCH_2–CH_2–OCH_2– in PTMC).

2.3.3. Synthesis of mPEG$_{43}$-*b*-P[(TMC-C)$_{20-x}$-HTMC$_x$-TMC$_{20}$]

The chiral monomer C (containing a –COOH group) was dissolved in dichloromethane and added to a 250 mL three-necked flask with a magnetic stir bar. Then, *N,N′*-dicyclohexylcarbodiimide (DCC) and 4-dimethylaminopyridine (DMAP) were dissolved in dichloromethane, and added dropwise to the mixture. After stirring for 0.5 h, mPEG$_{43}$-*b*-P(HTMC$_{20}$-TMC$_{20}$) (containing –OH groups) dissolved in dichloromethane was added dropwise to the above mixture. The reaction mixture was stirred for 72 h at room temperature. The fabrication of mPEG$_{43}$-*b*-P(HTMC$_{20}$-TMC$_{20}$) and the monomer C is shown in Table 1. The resulting mixture was washed with water and *N,N′*-dicyclohexyl urea was precipitated and filtered off. The filtrate was evaporated and concentrated. The crude product was purified by dissolving it in dichloromethane and precipitated in methanol. The precipitate was dried in a vacuum until constant weight was obtained.

IR (KBr, cm^{-1}): 2951, 2868($-CH_2-$, $-CH_3$); 1745 (C=O); 1674 (C=C); 1256, 1171 (C–O–C); ^{1}H NMR (δ, $CDCl_3$, 600 MHz): δ = 5.37 (–*CH*=C in cholesteryl), δ = 5.27 (–OCH_2–*CH*–CH_2O– in PHTMC), δ = 4.60 (–COOCH< in cholesteryl), δ = 4.36−4.25 (–OCH_2–CH–CH_2O), δ = 3.64 (–OCH_2– in mPEG), δ = 2.37−0.68 (the rest of the protons from PTMC, cholesteryl and –COO(CH_2)$_4$COO–).

Table 1. Fabrication of mPEG$_{43}$-*b*-P[(TMC-C)$_{20-x}$-HTMC$_x$-TMC$_y$] and 6-cholesteroxy-6-oxocaproic acid.

Copolymer	*m* (–OH)	*n* (–OH)	*m* (–COOH)	*n* (–COOH)
	g	mol	g	mol
*m*PEG$_{43}$-*b*-P[(TMC-C)$_{20}$-TMC$_{20}$]	5.00	0.016	8.28	0.016
*m*PEG$_{43}$-*b*-P[(TMC-C)$_{15}$-HTMC$_5$-TMC$_{20}$]	5.00	0.016	6.21	0.012
*m*PEG$_{43}$-*b*-P[(TMC-C)$_{12}$-HTMC$_8$-TMC$_{20}$]	5.00	0.016	3.58	0.008

2.4. Nanoprecipitation of Copolymers

The three amphiphilic LC block copolymers mPEG$_{43}$-*b*-P[(TMC-C)$_{20-x}$-HTMC$_x$-TMC$_y$] have unique features, with mPEG as a hydrophilic segment and P[(TMC-C)$_{20-x}$-HTMC$_x$-TMC$_y$] chains as hydrophobic segments. The micelles were prepared as follows: Firstly, 1.25 mg of each copolymer obtained in this study was dissolved in 5 mL of THF and stirred for 12 h. Secondly, the deionized water was added dropwise to the solution under slight shaking, and then the solution was placed for 5 min to balance. After that, the light transmittance T% of the mixed solution was tested by UV–vis. When the T% was stable, the mixed solution was transferred to a dialysis membrane (MWCO 3500Da) to remove residual THF. The solution was dialyzed in deionized water with corresponding different pH values for 72 h. The particle size and morphology of the micelles were determined by DLS and SEM.

2.5. Preparation of DOX-Loaded Micelles

100 mg of the LC copolymers were dissolved in 20 mL of THF and put in a 50-mL flask. Then, 10 mg of DOX·HCl and 10 uL TEA were added. The mixed solution was stirred for 30 min at room temperature. After that, 20 mL of phosphate buffer was added dropwise and stirred for 2 h. Afterward, the excess drug and residual THF were removed by dialysis (MWCO 3500Da) for 24 h to obtain a DOX-loaded microsphere aqueous solution. The chemical structure of DOX is shown in Scheme 2. Next, the drug loading (DL%) and the entrapment efficiency (EE%) were calculated according to the following formulas [37]:

$$DL\% = [(C_T - C_U)/C_L] \times 100\% \quad (1)$$

$$EE\% = (1 - C_U/C_T) \times 100\% \quad (2)$$

where C_U is the weight of free unloaded drug, C_T is the total weight of drug added to the system, and C_L is the total weight of copolymer micelles. The specific process was as follows: part of the DOX-loaded microsphere aqueous solution was taken to freeze-dry at −80 °C, and lyophilized DOX-loaded micelles were dissolved in THF/DMSO (1:1, *v*/*v*), in which the structure of the micelles would be destroyed and the drug would be released completely. The transmittance (T%) of the mixed solution was monitored with UV–vis, and the drug content ($C_T - C_U$) could be calculated based on the standard curve and the proportion of samples taken. In addition, C_T and C_L were the weights of DOX and the LC copolymer added to prepare DOX-loaded micelles, respectively.

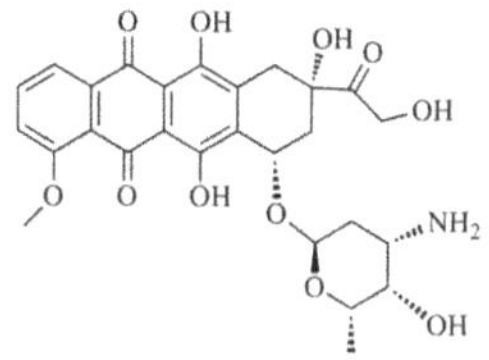

Scheme 2. The chemical structure of DOX.

2.6. In Vitro Drug Release

DOX-release study was carried out in thermostat oscillator (37 °C, 80 rpm/min) in phosphate buffers of different pH, respectively: (a) pH = 6.4; (b) pH = 7.4; (c) pH = 8.4. In brief, 10 mL of DOX-loaded micelle aqueous solution was placed in a dialysis membrane (MWCO 3500Da). Then, the dialysis membrane was immersed into the abovementioned media and shaken at 37 °C. 5 mL of buffer sample was taken out periodically and 5 mL of fresh buffer was added, and the in vitro DOX release was analyzed using a UV–vis spectrometer at the absorbance of 481 nm.

3. Results and Discussion

3.1. Thermal Stability

The ^{1}H NMR spectra of the three copolymers are showed in Figure S1. Figure 1 shows TGA curves of the block copolymers. The corresponding data of thermal decomposition and weight loss are summarized in Table 2.

In general, the decomposition temperature (T_d) is the temperature at which 5% weight loss of copolymers occurred. According to Table 2, compared with mPEG$_{43}$-*b*-P(BTMC$_{20}$-TMC$_{20}$), the T_d of *m*PEG$_{43}$-*b*-P(HTMC$_{20}$-TMC$_{20}$) decreased by 92.3 °C. The main reason for this was that the existence of benzene groups enhanced intermolecular π–π conjugation so that the thermal stability of the copolymer increased. At the same time, the hydroxyl groups were more active and easily resulted in the occurrence of thermal decomposition. For the LC copolymers, when bulky cholesteryl units were introduced into side chains of the polycarbonate, the corresponding T_d increased. Furthermore, with a higher content of LC units, the LC copolymer was more stable. For example, *m*PEG$_{43}$-*b*-P[(TMC-C)$_{20}$-TMC$_{20}$] containing the highest content of LC units revealed the highest T_d and the best stability. It suggested that the existence of LC units may cause a strong interaction between the repeating units and the polymer chains, which would enhance the stability of the copolymers.

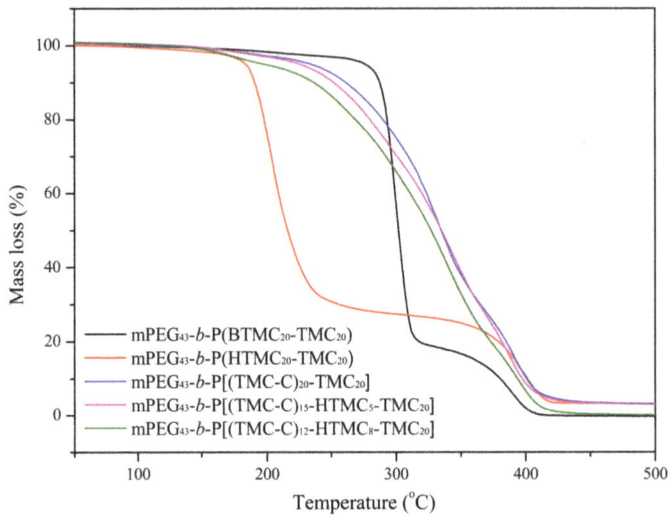

Figure 1. TGA curves of the copolymers.

Table 2. TGA data of the copolymers.

Copolymer	T_d [a] (°C)	Weight Loss (%)				
		200 °C	250 °C	300 °C	350 °C	400 °C
*m*PEG$_{43}$-*b*-P(BTMC$_{20}$-TMC$_{20}$)	276.5	2.04	3.40	42.83	83.68	98.35
*m*PEG$_{43}$-*b*-P(HTMC$_{20}$-TMC$_{20}$)	184.2	23.19	69.22	72.63	75.04	89.32
*m*PEG$_{43}$-*b*-P[(TMC-C)$_{20}$-TMC$_{20}$]	234.4	3.31	7.79	25.31	62.36	90.09
*m*PEG$_{43}$-*b*-P[(TMC-C)$_{15}$-HTMC$_5$-TMC$_{20}$]	225.8	3.45	9.74	29.86	61.31	91.62
*m*PEG$_{43}$-*b*-P[(TMC-C)$_{12}$-HTMC$_8$-TMC$_{20}$]	197.2	6.15	14.33	34.86	67.91	94.21

3.2. Liquid Crystal Behavior

The LC behavior of the copolymers was investigated with DSC, POM, and XRD. The corresponding glass transition temperatures (T_g) obtained during the second heating cycles from DSC and the isotropic temperature (T_i) obtained from POM are summarized in Table 3. Typical DSC curves of five copolymers are shown in Figure 2.

Table 3. Thermal properties of the copolymers.

Copolymer	Phase Transition Temperature (°C)
*m*PEG$_{43}$-*b*-P(BTMC$_{20}$-TMC$_{20}$)	g −25.0 I
*m*PEG$_{43}$-*b*-P(HTMC$_{20}$-TMC$_{20}$)	g −24.1 I
*m*PEG$_{43}$-*b*-P[(TMC-C)$_{20}$-TMC$_{20}$]	g −4.6 SmA 180.0 I
*m*PEG$_{43}$-*b*-P[(TMC-C)$_{15}$-HTMC$_5$-TMC$_{20}$]	g −7.4 SmA 169.8 I
*m*PEG$_{43}$-*b*-P[(TMC-C)$_{12}$-HTMC$_8$-TMC$_{20}$]	g −11.2 SmA 142.6 I

g: glass state; SmA: smectic A phase; I: isotropic.

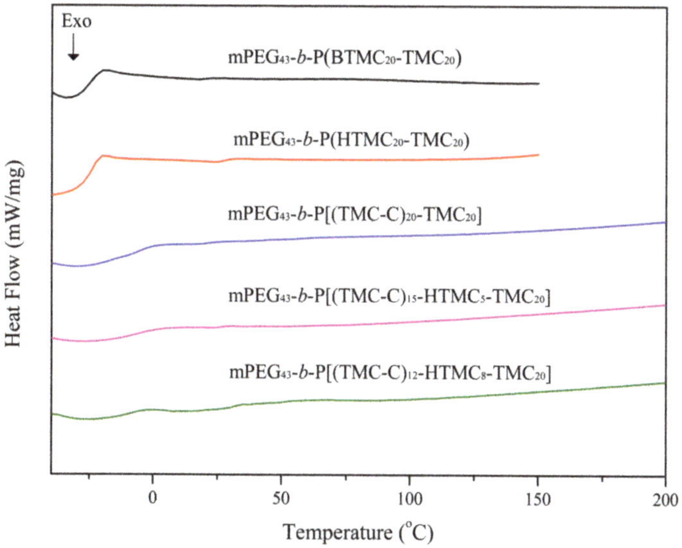

Figure 2. DSC curves of the copolymers.

In general, the thermal properties of side-chain polymers mainly depend on the polymer backbone and the nature of the side groups. For all the copolymers with the same polycarbonate backbone in this investigation, the side groups had a significant influence on the glass transition temperature of the copolymers. As shown in Table 3, the corresponding T_g increased when strong-polarity hydroxyl groups were introduced into the polycarbonate backbone. In addition, when cholesteryl units were introduced, the corresponding T_g also increased because the copolymers had a higher average molecular weight and more rigid side-chain cholesteryl groups. As seen in Figure 2, the copolymer mPEG$_{43}$-*b*-P(HTMC$_{20}$-TMC$_{20}$) with side hydroxyl groups showed T_g at −24.1 °C, while the three LC copolymers with side-chain cholesteryl groups mPEG$_{43}$-*b*-P[(TMC-C)$_{12}$-HTMC$_8$-TMC$_{20}$], mPEG$_{43}$-*b*-P[(TMC-C)$_{15}$-HTMC$_5$-TMC$_{20}$], and mPEG$_{43}$-*b*-P[(TMC-C)$_{20}$-TMC$_{20}$] showed T_g at −11.2 °C, −7.4 °C, and −4.6 °C, respectively. It suggested that the copolymers containing LC units exhibited LC phases at human body temperature. Therefore, they may be used clinically as self-assembling materials with orientational order.

For the LC polymers, the formation of the mesophase may be influenced by the polymer backbone, aspect ratio of the mesogen, flexible spacer, and polarity of the molecules [40,41]. Optical textures of the three LC copolymers are shown in Figure 3. POM showed that all the LC copolymers exhibited the fan-shaped texture of a smectic A (SmA) phase in heating and cooling processes. According to the results reported by Hu et al. [29], the corresponding chiral monomer C showed a cholesteric phase. The result indicated that the macromolecular chain might hinder the formation of a cholesteric helical supermolecular structure and an ordered organization into the mesophase.

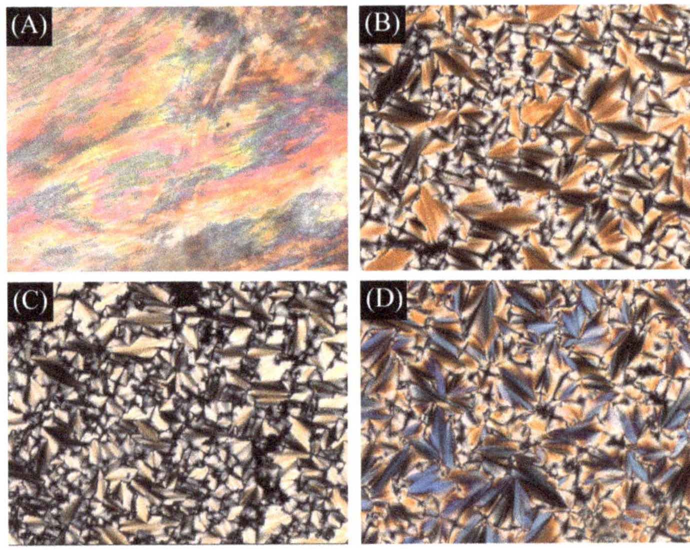

Figure 3. Polarizing optical microscope (POM) texture of the LC copolymers (200×). Polarizing optical microscope (POM) texture of the LC copolymers. (**A**) LC texture of a SmA phase at 40 °C for mPEG$_{43}$-*b*-P[(TMC-C)$_{20}$-TMC$_{20}$]; (**B**) fan-shaped texture of a SmA phase at 116 °C for mPEG$_{43}$-*b*-P[(TMC-C)$_{20}$-TMC$_{20}$]; (**C**) fan-shaped texture of a SmA phase at 120 °C for mPEG$_{43}$-*b*-P[(TMC-C)$_{15}$-HTMC$_5$-TMC$_{20}$]; (**D**) fan-shaped texture of a SmA phase at 104 °C for mPEG$_{43}$-*b*-P[(TMC-C)$_{12}$-HTMC$_8$-TMC$_{20}$].

XRD is a powerful instrument for the identification of mesophase structure. XRD measurements of the LC copolymers were carried out at the mesophase temperature of 120 °C. As an example, the XRD pattern of mPEG$_{43}$-*b*-P[(TMC-C)$_{20}$-TMC$_{20}$] is shown in Figure 4. The XRD pattern showed a small-angle reflection and a broad diffuse peak at a wide angle. In general, a sharp reflection in the small-angle region, corresponding to the periodic distance, is characteristic of a smectic phase. Combined with POM texture, the LC copolymers obtained in this study can be judged as having a SmA phase. According to Figure 4, the molecular layer spacing d was calculated to be 51.97 Å. The all-trans molecular length L of the most extended conformation of mPEG$_{43}$-*b*-P[(TMC-C)$_{20}$-TMC$_{20}$] was about 26.3 Å, which could be calculated by using ChemBio3D-Ultra and MM2 minimal energy parameters. A d/L ratio of 1.97 ($d \approx 2L$) was calculated, indicating an orderly doubled arrangement of mPEG$_{43}$-*b*-P[(TMC-C)$_{20}$-TMC$_{20}$]. This partial bilayer structure was similar to that of a SmA$_d$ phase formed by polar mesogens. Similar results also have been reported [42,43]. The possible molecular arrangement model of the doubled SmA layer for mPEG$_{43}$-*b*-P[(TMC-C)$_{20}$-TMC$_{20}$] is shown in Figure 5.

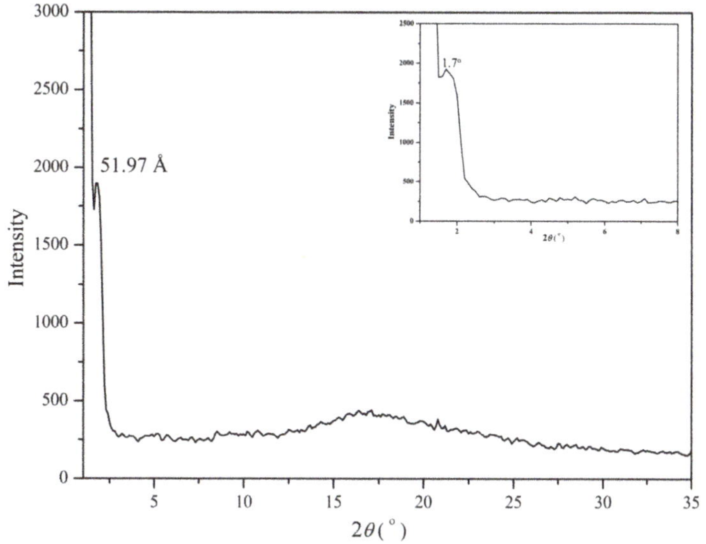

Figure 4. X-ray diffraction (XRD) pattern of mPEG$_{43}$-*b*-P[(TMC-C)$_{20}$-TMC$_{20}$] at 120 °C.

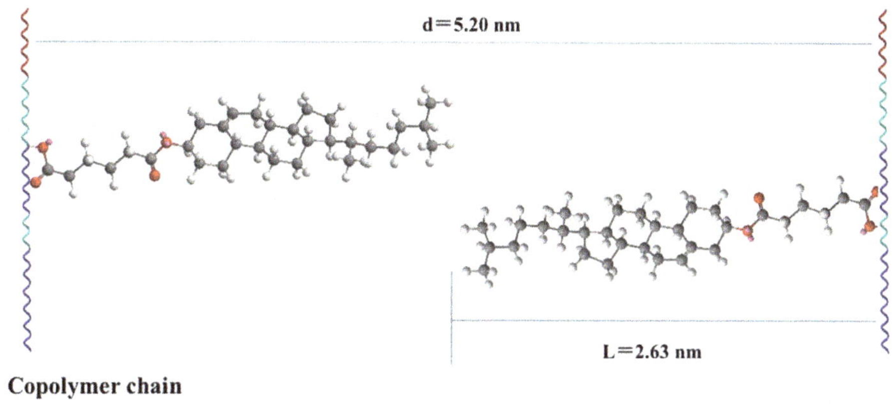

Figure 5. Schematic representation of the doubled SmA layer arrangements of mPEG$_{43}$-*b*-P[(TMC-C)$_{20}$-TMC$_{20}$].

3.3. *Self-Assembly Behavior*

To research the self-assembly behavior of the LC copolymers, the classic nanoprecipitated self-assembly or dialysis method is used with different acidic environments. The block copolymer mPEG$_{43}$-*b*-P[(TMC-C)$_{20}$-TMC$_{20}$] was first dissolved in THF in five vials equipped with a sealing cap, and then aqueous solutions with different pH values were added dropwise, respectively. The light transmittance of mixture solutions with different water content was determined by UV–vis at 267 nm, and the corresponding relationships are shown in Figure 6.

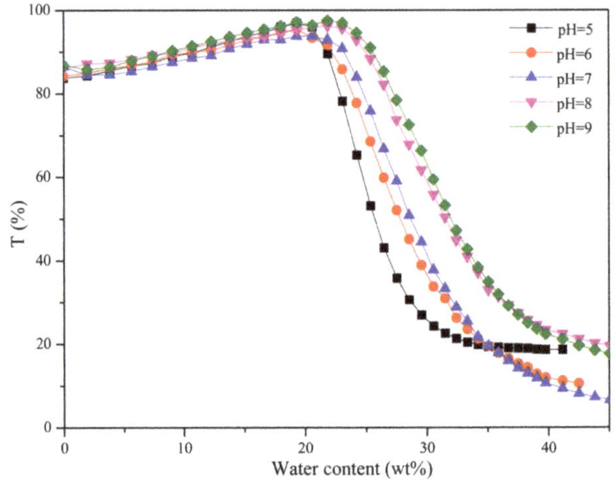

Figure 6. Relationships between light transmission at 267 nm and water content of mPEG$_{43}$-*b*-P[(TMC-C)$_{20}$-TMC$_{20}$] for the self-assembly process at different pH levels.

When the water content was less than 20 wt %, the mixed solution remained transparent and the light transmittance changed a little. With the increase of water content, the light transmittance of the mixed solution revealed a slowly reduced trend, suggesting that the copolymer began to form micelles. The light transmittance was stable when the water content reached 40 wt %, indicating that the formation of micelles was nearly finished. Figure 6 shows that the self-assembly process speeded up gradually with the decrease of the pH value. There were hydrogen bonding forces present at pH = 7.0, and the hydrophilia of kernels was enhanced due to the protonation of hydroxyl groups in a weak-acid condition, and both of these would accelerate the self-assembly process.

After the formation of stable aggregates, the mixture solutions were dialyzed for 72 h under room temperature to remove THF completely, and then the particle size and distribution of the self-assembled micelles were characterized by DLS; the results are shown in Table 4. Particle dispersion of the self-assembled micelles was about 0.1–0.2 at pH 5–9, indicating that the size distribution was relatively uniform. When pH > 7, the size of particle changed a little because the deprotonation of the –OH groups in copolymer would occur and result in the disruption of hydrogen bonds and the enhancement of the hydrophobicity of the copolymer. As the pH value decreased, the particle size increased. These results showed that the hydrophilia of kernels was enhanced because of the protonation of hydroxyl groups. Moreover, the mutual exclusion of the proton segment with stronger positive charge may promote the micellar swelling and further result in the increase of particle size.

Table 4. Self-assembled particle size distribution and PDI of micelles at different pH levels.

Copolymer	pH	Z-Average, d/nm	PDI
P1	5	346.3	0.136
P1	6	307.4	0.123
P1	7	296.4	0.146
P1	8	295.1	0.149
P1	9	301.7	0.179
P2	7	313.4	0.132
P3	7	352.9	0.067

P1: mPEG$_{43}$-*b*-P[(TMC-C)$_{20}$-TMC$_{20}$]; P2: mPEG$_{43}$-*b*-P[(TMC-C)$_{15}$-HTMC$_5$-TMC$_{20}$]; P3: mPEG$_{43}$-*b*-P[(TMC-C)$_{12}$-HTMC$_8$-TMC$_{20}$]; PDI: the distribution of the particle size.

Three LC copolymers were dissolved in THF and assembled in the deionized water at pH = 7. The relationships between light transmittance and the water content of mixture solution were determined by UV–vis and shown in Figure 7. The LC copolymers *m*PEG_{43}-*b*-P[(TMC-C)$_{20}$-TMC_{20}] and *m*PEG_{43}-*b*-P[(TMC-C)$_{15}$-$HTMC_5$-TMC_{20}] began to assemble when the water content was about 20 wt %, and became stable as the water content reached 41 wt % and 46 wt %, respectively. As for *m*PEG_{43}-*b*-P[(TMC-C)$_{12}$-$HTMC_8$-TMC_{20}], which contained fewer LC units, it began to assemble at a water content of 45 wt %, and remained stable until the water content reached 59 wt %. Therefore, the LC content played an important role in the self-assembly behavior. The higher the LC content, the faster the self-assembly process was. It was because of the stronger driving force for self-assembly caused by the improved orientation ability of copolymers as the LC content increased.

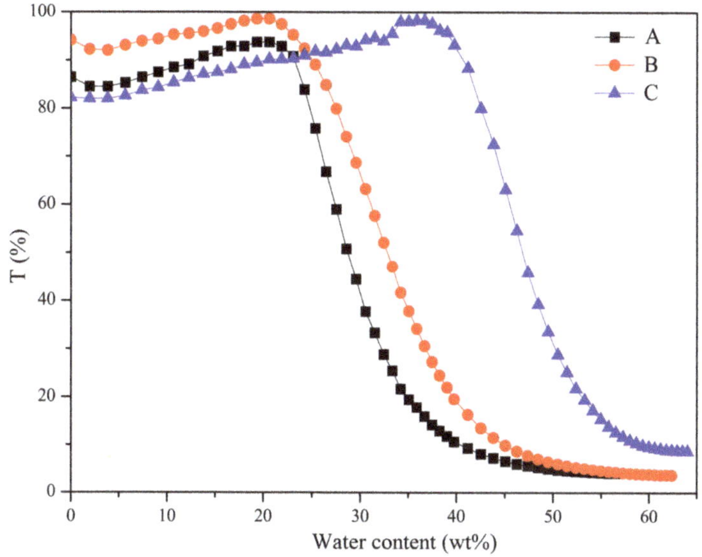

Figure 7. Relationships between light transmission at 267 nm and water content for the self-assembly behavior of the different copolymer structures (A) *m*PEG_{43}-*b*-P[(TMC-C)$_{20}$ TMC_{20}]; (B) *m*PEG_{43}-*b*-P[(TMC-C)$_{15}$-$HTMC_5$-TMC_{20}]; and (C) mPEG_{43}-*b*-P[(TMC-C)$_{12}$-$HTMC_8$-TMC_{20}].

As shown in Table 4, the aggregate size of the LC block copolymers was smaller with increasing LC unit content. The introduction of LC units would lead to the decrease of hydroxyl groups in the copolymer, and the copolymers became more hydrophobic. In addition, as the LC unit content increased, the orientation of the molecule would be enhanced to provide a stronger driving force for self-assembly, and produce a more compact structure, resulting in smaller particle size.

The morphologies of the *m*PEG_{43}-*b*-P[(TMC-C)$_{20}$-TMC_{20}] micelles were determined by SEM, and presented in Figure 8. The results indicated that the aggregates exhibited spherical micelles. The LC copolymers were dissolved in THF solution and the chain was stretched. When the selective solvent of water was added, the polycarbonates with LC side segments quickly froze due to the hydrophobicity, and THF was pumped continuously; as a result the micelles formed gradually. However, in this process, the micelle structure was likely to continue adjusting because the molecules in the LC state possessed orientation ability with the addition of water. Thus, the copolymer made further arrangements neatly and the inner space of the hydrophobic kernel was more compact, and a solid sphere aggregate formed, finally. Since there was a certain liquidity of LCs, microsphere fusion appeared, as shown in Figure 8.

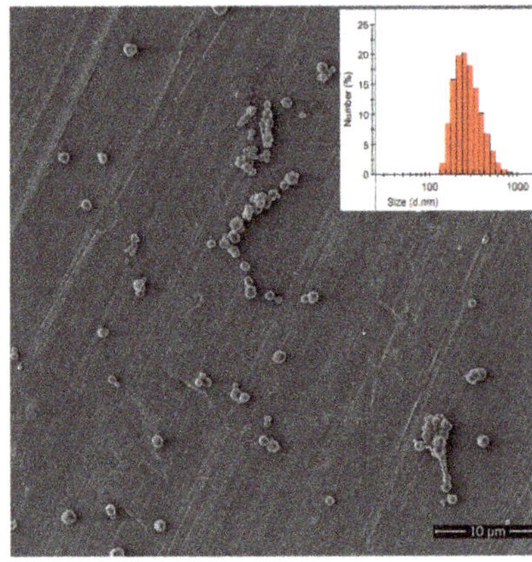

Figure 8. Scanning electron microscope (SEM) image of the self-assembled aggregates for *m*PEG$_{43}$-*b*-P[(TMC-C)$_{20}$ TMC$_{20}$] in pH = 7.

3.4. DOX-Loaded Micelles

Figure 9 depicts particle size and distribution situation of DOX-loaded micelles characterized by DLS, and the detailed data are listed in Table 5.

The copolymer structure and the pH value had a critical influence on the DOX-loading efficiency and the particle size of the drug-loaded micelles. As shown in Figure 9 and Table 5, the particle size of DOX-loaded *m*PEG$_{43}$-*b*-P[(TMC-C)$_{12}$-HTMC$_{8}$-TMC$_{20}$] tended to form double peaks in the self-assembly process, and the size of DOX-loaded *m*PEG$_{43}$-*b*-P[(TMC-C)$_{12}$-HTMC$_{8}$-TMC$_{20}$] was larger than that of other copolymers. It was ascribed to the lowest content of hydrophobic LC units of *m*PEG$_{43}$-*b*-P[(TMC-C)$_{12}$-HTMC$_{8}$-TMC$_{20}$] making it the most hydrophilic, resulting in the swelling of the corresponding microspheres. Furthermore, the size of DOX-loaded *m*PEG$_{43}$-*b*-P[(TMC-C)$_{12}$-HTMC$_{8}$-TMC$_{20}$] micelles was larger at pH = 7.4 than that at pH = 6.4 and pH = 8.4 because of the maximum drug loading. However, the size of DOX-loaded *m*PEG$_{43}$-*b*-P[(TMC-C)$_{15}$-HTMC$_{5}$-TMC$_{20}$] micelles was minimum at pH = 7.4. Here are some reasons: Firstly, the strong hydrogen bond force between the copolymers and DOX-loaded microspheres caused them to shrink closely together. Secondly, the copolymer *m*PEG$_{43}$-*b*-P[(TMC-C)$_{15}$-HTMC$_{5}$-TMC$_{20}$], with more LC units and its hydrophobic core, arranged neatly and packed tightly so as to form a smaller particle size of microspheres. However, when the content of LC units in the copolymers continued to increase, the particle size also increased, owing to the increase of molecular volume. This is why the particle size of DOX-loaded *m*PEG$_{43}$-*b*-P[(TMC-C)$_{20}$-TMC$_{20}$] (256.6 nm) at pH = 7.4 was larger than that of DOX-loaded *m*PEG$_{43}$-*b*-P[(TMC-C)$_{15}$-HTMC$_{5}$-TMC$_{20}$] (225.2 nm).

Compared to other copolymers, mPEG$_{43}$-*b*-P[(TMC-C)$_{12}$-HTMC$_{8}$-TMC$_{20}$] showed the lowest DOX-loading efficiency under the conditions of pH $\leq$ 7.4. As the most hydrophilic copolymer, it was difficult to load more hydrophobic drug, such as DOX.

Table 5. Particle size and the DOX-loading efficiency for the drug-loaded copolymer micelles.

Copolymer	P1			P2			P3		
	pH = 6.4	pH = 7.4	pH = 8.4	pH = 6.4	pH = 7.4	pH = 8.4	pH = 6.4	pH = 7.4	pH = 8.4
Size (nm)	219.0	256.6	304.5	286.3	225.2	298.2	319.6	398.6	305.6
PDI	0.112	0.224	0.506	0.083	0.071	0.312	0.322	0.428	0.253
DL (%)	0.544	0.541	0.226	0.484	0.572	0.343	0.447	0.508	0.360
EE (%)	33.21	31.82	13.20	32.92	32.62	20.74	28.92	27.62	21.72

P1: mPEG$_{43}$-*b*-P[(TMC-C)$_{20}$-TMC$_{20}$]; P2: mPEG$_{43}$-*b*-P[(TMC-C)$_{15}$-HTMC$_{5}$-TMC$_{20}$]; P3: mPEG$_{43}$-*b*-P[(TMC-C)$_{12}$-HTMC$_{8}$-TMC$_{20}$]; DL: drug loading efficiency; EE: entrapment efficiency.

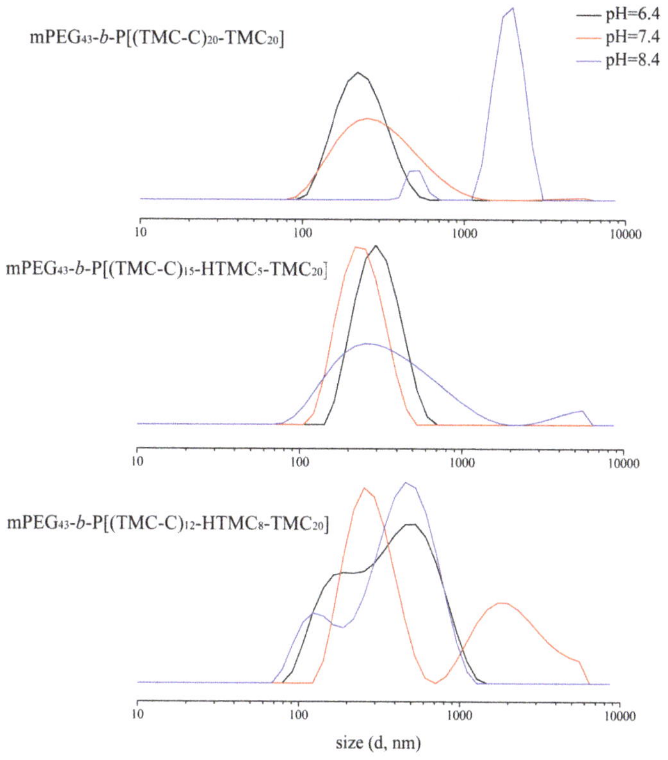

Figure 9. pH-responsive particle sizes of the DOX-loaded copolymer micelles at 37 °C in phosphate buffer (pH = 6.4, 7.4, 8.4) by DLS.

The DOX loading of all copolymer micelles reached the maximum value when pH = 7.4, owing to the hydrogen bond force between copolymers and DOX being strongest at pH = 7.4. Furthermore, the drug loading of the three copolymers all exhibited a decreasing trend in an alkaline condition. The results showed that DOX had a low solubility under an alkaline condition, and the copolymer was ionized which gave it a certain repulsion towards DOX, causing less DOX to be loaded into the micelle core finally.

The morphologies of DOX-loaded copolymer micelles were characterized by SEM. The morphology of *m*PEG$_{43}$-*b*-P[(TMC-C)$_{20}$-TMC$_{20}$] micelles is typical, as shown in Figure 10. The drug-loaded copolymers showed spherical and irregular shapes, which was similar to micelles without DOX. However, it was surprising that the size of the micelles showed a decreased trend after loading of the drug. The reason may be that strong hydrogen bond forces between the copolymers and DOX shrink the particle size.

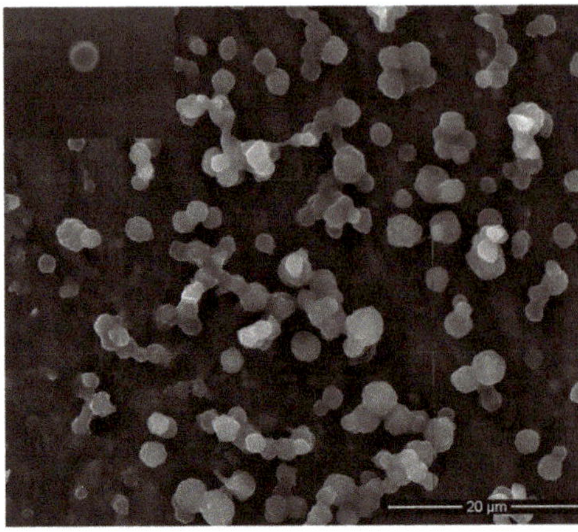

Figure 10. SEM image of DOX-loaded mPEG$_{43}$-*b*-P[(TMC-C)$_{20}$-TMC$_{20}$] micelles.

3.5. *In Vitro DOX Release from the Micelles*

In vitro DOX release profiles of the copolymers *m*PEG$_{43}$-*b*-P[(TMC-C)$_{20}$-TMC$_{20}$], *m*PEG$_{43}$-*b*-P[(TMC-C)$_{15}$-HTMC$_{5}$-TMC$_{20}$], and *m*PEG$_{43}$-*b*-P[(TMC-C)$_{12}$-HTMC$_{8}$-TMC$_{20}$] in various media are shown in Figure 11. And the simulation diagram of DOX loading and realease behavior is showed in Figure S2.

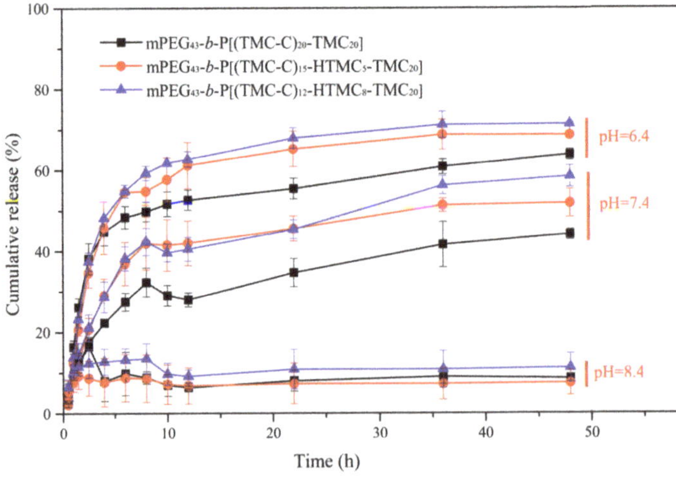

Figure 11. DOX release profiles of the LC copolymer micelles at 37 °C in PBS at different pH levels.

At pH = 8.4, the drug within the three drug-loaded copolymer micelles was slowly released, and only 10% DOX was released in the end. At pH = 7.4, the DOX release rate was obviously accelerated, and its cumulative release could reach 40–60% within 48 h. The cumulative DOX release could increase up to 60–70% when the pH was decreased to 6.4. For *m*PEG$_{43}$-*b*-P[(TMC-C)$_{20}$-TMC$_{20}$], 63%, 44%, and 8% DOX was released within 48 h at pH = 6.4, 7.4, and 8.4, respectively. The results could be explained as due to the fact that the molecular protonation extent was stronger under acidic conditions, and the water-solubility of DOX and copolymers increased as well, so ultimately, the DOX release rate was accelerated.

The cumulative DOX release at pH = 7.4 for the drug micelles of *m*PEG$_{43}$-*b*-P[(TMC-C)$_{15}$-HTMC$_5$-TMC$_{20}$] and mPEG$_{43}$-*b*-P[(TMC-C)$_{12}$-HTMC$_8$-TMC$_{20}$] was 51% and 58%, respectively. The experimental results showed that the existence of LC units could reduce the release rate of DOX to some extent.

4. Conclusions

In summary, three new LC amphipathic polycarbonate copolymers containing side-chain cholesteryl units were synthesized and characterized. The LC copolymers all showed high thermal stability, exhibited a fan-shape texture, and revealed a double mesogenic molecular arrangement of a SmA phase. In addition, with the increase of LC unit content in the copolymers, the corresponding T_m, T_i, and T_d increased. In the process of self-assembly, the aggregates of the LC copolymers exhibited spherical micelles with pH responsiveness. Moreover, the copolymers containing higher contents of LC units tended to form smaller microspheres. The LC copolymers were taken as carriers in which to load DOX, and study the drug loading capacity of the copolymers and in vitro drug release of micelles. The drug loading capacity was best at pH = 7.4. The size of the drug-loaded copolymer microspheres was not only related to the pH value, but also related to the contents of LC units in the copolymers. The drug release process also had a pH response, and DOX could be released quickly in a weakly acidic environment, which is expected to enable it to be used clinically for applications in drug delivery to specific tumor sites.

Supplementary Materials: The following are available online at www.mdpi.com/2079-4991/8/4/195/s1. Synthesis and structural characterization: Figure S1: ^{1}H NMR spectra of LC copolymers. A: mPEG$_{43}$-*b*-P[(TMC-C)$_{20}$-TMC$_{20}$]; B: mPEG$_{43}$-*b*-P[(TMC-C)$_{15}$-HTMC$_5$-TMC$_{20}$]; C: mPEG$_{43}$-*b*-P[(TMC-C)$_{12}$-HTMC$_8$-TMC$_{20}$]; Figure S2: The simulation diagram of DOX loading and release behavior.

Acknowledgments: This work was supported by the National Key R&D Program of China (2016YFC1000902), the National Natural Science Foundation of China (21702026 and 51503093), the Fundamental Research Funds for the Central Universities (N160504001 and N160503001), the Science and Technology Foundation of Liaoning Province (201501116 and 20170520185), and the Science and Technology Bureau of Shenyang (F16-205-1-37, RC170359).

Author Contributions: Jianshe Hu and Liqun Yang conceived and designed the experiments; Yujiao Xie performed the experiments, analyzed the data, and wrote an earlier draft; Xiao-Feng Liu helped analyze the data and polish the paper; Zhuang Hu, Zhipeng Hou, and Zhihao Guo helped perform the experiments; Zhangpei Chen helped analyze the data.

Conflicts of Interest: No potential conflict of interest was reported by the authors.

References

1. Dicastillo, L.C.; Garrido, L.; Alvarado, N.; Romero, J.; Palma, J.L.; Galotto, M.J. Improvement of polylactide properties through cellulose nanocrystals embedded in poly(vinyl alcohol) electrospun nanofibers. *Nanomaterials* **2017**, *7*, 106. [CrossRef] [PubMed]
2. Wang, H.B.; Wang, Y.; Chen, Y.J.; Jin, Q.; Ji, J. A biomimic pH-sensitive polymeric prodrug based on polycarbonate for intracellular drug delivery. *Polym. Chem.* **2014**, *5*, 854–861. [CrossRef]
3. Ishii, D.; Ying, T.H.; Mahara, A.; Murakami, S.; Yamaoka, T. In vivo tissue response and degradation behavior of PLLA and stereocomplexed PLA nanofibers. *Biomacromolecules* **2009**, *10*, 237–242. [CrossRef] [PubMed]
4. Wang, Y.C.; Li, P.W.; Truong-Dinh Tran, T.; Zhang, J.; Kong, L.X. Manufacturing techniques and surface engineering of polymer based nanoparticles for targeted drug delivery to cancer. *Nanomaterials* **2016**, *6*, 26. [CrossRef] [PubMed]
5. Solovieva, A.; Miroshnichenko, S.; Kovalskii, A.; Permyakova, E.; Popov, Z.; Dvorakova, E.; Kiryukhantsev-Korneev, P.; Obrosov, A.; Polcak, J.; Zajickova, L.; et al. Immobilization of platelet-rich plasma onto COOH plasma-coated PCL nanofibers boost viability and proliferation of human mesenchymal stem cells. *Polymers* **2017**, *9*, 736. [CrossRef]
6. Liu, X.F.; Xu, X.X.; Li, Q.; Hu, J.S.; Yang, L.Q.; Chen, Q.F.; Lu, Y.F. New side chain cholesterol-functionalised aliphatic polycarbonate copolymer: Synthesis and phase behaviour. *Liq. Cryst.* **2017**, *44*, 1–9. [CrossRef]

7. Zhang, Z.; Kuijer, R.; Bulstra, S.K.; Grijpma, D.W.; Feijen, J. The in vivo and in vitro degradation behavior of poly(trimethylene carbonate). *Biomaterials* **2006**, *27*, 1741–1748. [CrossRef] [PubMed]
8. Wang, X.L.; Zhuo, R.X.; Liu, L.J.; He, F.; Liu, G. Synthesis and characterization of novel aliphatic polycarbonates. *J. Polym. Sci. Part A Polym. Chem.* **2002**, *40*, 70–75. [CrossRef]
9. Wang, X.L.; Zhuo, R.X.; Huang, S.W.; Liu, L.J.; He, F. Synthesis, characterization and in vitro cytotoxicity of poly[(5-benzyloxy-trimethylene carbonate)-*co*-(trimethylene carbonate)]. *Macromol. Chem. Phys.* **2002**, *203*, 985–990. [CrossRef]
10. Zeng, F.Q.; Liu, J.B.; Allen, C. Synthesis and characterization of biodegradable poly(ethylene glycol)-block-poly(5-benzyloxy-trimethylene carbonate) copolymers for drug delivery. *Biomacromolecules* **2004**, *5*, 1810–1817. [CrossRef] [PubMed]
11. Grzybowski, B.A.; Wilmer, C.E.; Kim, J.; Browne, K.P.; Bishop, K.J.M. Self-assembly: From crystals to cells. *Soft Matter* **2009**, *5*, 1110–1128. [CrossRef]
12. Han, L.; Ma, H.W.; Zhu, S.Q.; Liu, P.B.; Shen, H.Y.; Yang, L.C.; Tan, R.; Huang, W.; Li, Y. Effect of topology and composition on liquid crystal order and self-assembly performances driven by asynchronously controlled grafting density. *Macromolecules* **2017**, *50*, 8334–8345. [CrossRef]
13. Zhou, J.S.; Dong, Y.C.; Zhang, Y.Y.; Liu, D.S.; Yang, Z.Q. The assembly of DNA amphiphiles at liquid crystal-aqueous interface. *Nanomaterials* **2016**, *6*, 229. [CrossRef] [PubMed]
14. Cano-Sarabia, M.; Angelova, A.; Ventosa, N.; Lesieur, S.; Veciana, J. Cholesterol induced CTAB micelle-to-vesicle phase transitions. *J. Colloid Interface Sci.* **2010**, *350*, 10–15. [CrossRef] [PubMed]
15. Zerkoune, L.; Lesieur, S.; Putaux, J.L.; Choisnard, L.; Geze, A.; Wouessidjewe, D.; Angelov, B.; Vebert-Nardin, C.; Doutch, J.; Angelona, A. Mesoporous self-assembled nanoparticles of biotransesterified cyclodextrins and nonlamellar lipids as carriers of water-insoluble substances. *Soft Matter* **2016**, *12*, 7539–7550. [CrossRef] [PubMed]
16. Zhang, J.H.; Bazuin, C.G.; Freiberg, S.; Brisse, F.; Zhu, X.X. Effect of side chain structure on the liquid crystalline properties of polymers bearing cholesterol, dihydrocholesterol and bile acid pendant groups. *Polymer* **2005**, *46*, 7266–7272. [CrossRef]
17. Zhou, Y.X.; Briand, V.A.; Sharma, N.; Ahn, S.K.; Kasi, R.M. Polymers comprising cholesterol: Synthesis, self-assembly, and applications. *Materials* **2009**, *2*, 636–660. [CrossRef]
18. Ahn, S.K.; Le, L.T.N.; Kasi, R.M. Synthesis and characterization of side-chain liquid crystalline polymers bearing cholesterol mesogen. *J. Polym. Sci. Part A Polym. Chem.* **2009**, *47*, 2690–2701. [CrossRef]
19. Nagahama, K.; Ueda, Y.; Ouchi, T.; Ohya, Y. Exhibition of soft and tenacious characteristics based on liquid crystal formation by introduction of cholesterol groups on biodegradable lactide copolymer. *Biomacromolecules* **2007**, *8*, 3938–3943. [CrossRef] [PubMed]
20. Zhou, F.; Zhang, Z.B.; Jiang, G.Q.; Liu, J.J.; Chen, X.F.; Li, Y.W.; Zhou, N.C.; Zhu, X.L. Self-assembly of amphiphilic macrocycles containing polymeric liquid crystal grafts in solution. *Polym. Chem.* **2016**, *7*, 2785–2789. [CrossRef]
21. Zhou, L.; Zhang, D.P.; Hocine, S.; Pilone, A.; Trepout, S.; Marco, S.; Thomas, C.; Guo, J.; Li, M.H. Transition from smectic nanofibers to smectic vesicles in the self-assemblies of PEG-*b*-liquid crystal polycarbonates. *Polym. Chem.* **2017**, *8*, 4776–4780. [CrossRef]
22. Venkataraman, S.; Lee, A.L.; Maune, H.T.; Hedrick, J.L.; Prabhu, V.M.; Yang, Y.Y. Formation of disk-and stacked-disk-like self-assembled morphologies from cholesterol-functionalized amphiphilic polycarbonate diblock copolymers. *Macromolecules* **2013**, *46*, 4839–4846. [CrossRef]
23. Wang, Z.; Luo, T.; Sheng, R.L.; Li, H.; Sun, J.J.; Cao, A.M. Amphiphilic diblock terpolymer PMAgala-*b*-P(MAA-*co*-MAChol)s with attached galactose and cholesterol grafts and Their intracellular pH-responsive doxorubicin delivery. *Biomacromolecules* **2015**, *17*, 98–110. [CrossRef] [PubMed]
24. Wan, T.; Zou, T.; Cheng, S.X.; Zhuo, R.X. Synthesis and characterization of biodegradable cholesteryl end-capped polycarbonates. *Biomacromolecules* **2005**, *6*, 524–529. [CrossRef] [PubMed]
25. Zou, T.; Cheng, S.X.; Zhuo, R.X. Synthesis and enzymatic degradation of end-functionalized biodegradable polyesters. *Colloid Polym. Sci.* **2005**, *283*, 1091–1099. [CrossRef]
26. Zhang, L.; Wang, Q.R.; Jiang, X.S.; Jiang, X.S.; Cheng, S.X.; Zhuo, R.X. Studies on functionalization of poly (ε-caprolactone) by a cholesteryl moiety. *J. Biomater. Sci. Polym. Ed.* **2005**, *16*, 1095–1108. [CrossRef] [PubMed]
27. Zou, T.; Li, F.; Cheng, S.X.; Zhuo, R.X. Synthesis and characterization of end-capped biodegradable oligo/poly(trimethylene carbonate)s. *J. Biomater. Sci. Polym. Ed.* **2006**, *17*, 1093–1106. [CrossRef]

28. Chen, G.X.; Yang, L.Q.; Liu, X.F.; Xie, Y.J.; Guo, Z.H.; Li, M.; Guo, J.; Hu, J.S. Main-chain biodegradable liquid crystal derived from cholesteryl derivative end-capped poly(trimethylene carbonate): Synthesis and characterisation. *Liq. Cryst.* **2017**, *44*, 1050–1058. [CrossRef]
29. Xu, X.X.; Liu, X.F.; Li, Q.; Hu, J.S.; Chen, Q.F.; Yang, L.Q.; Lu, Y.H. New amphiphilic polycarbonates with side functionalized cholesteryl groups as biomesogenic units: Synthesis, structure and liquid crystal behavior. *RSC Adv.* **2017**, *7*, 14176–14185. [CrossRef]
30. Qi, Y.P.; Miao, Z.M.; Cheng, S.X.; Zhang, X.Z. Fabrication and drug release properties of poly(5-benzyloxy-trimethylene-*co*-glycolide) microspheres. *J. Appl. Polym. Sci.* **2010**, *115*, 3451–3455. [CrossRef]
31. Han, J.Y.; Zhao, D.D.; Li, D.; Wang, X.H.; Jin, Z.; Zhao, K. Polymer-based nanomaterials and applications for vaccines and drugs. *Polymers* **2018**, *10*, 31. [CrossRef]
32. Angelova, A.; Garamus, V.M.; Angelov, B.; Tian, Z.F.; Li, Y.W.; Zou, A.H. Advances in structural design of lipid-based nanoparticle carriers for delivery of macromolecular drugs, phytochemicals and anti-tumor agents. *Adv. Colloid Interface Sci.* **2017**. [CrossRef] [PubMed]
33. Angelov, B.; Garamus, V.M.; Drechsler, M.; Angelova, A. Structural analysis of nanoparticulate carriers for encapsulation of macromolecular drugs. *J. Mol. Liq.* **2017**, *235*, 83–89. [CrossRef]
34. Angelova, A.; Angelov, B.; Mutafchieva, R.; Lesieur, S.; Couvreur, P. Self-assembled multicompartment liquid crystalline lipid carriers for protein, peptide, and nucleic acid drug delivery. *Acc. Chem. Res.* **2010**, *44*, 147–156. [CrossRef] [PubMed]
35. Angelov, B.; Angelova, A.; Filippov, S.K.; Drechsler, M.; Stepanek, P.; Lesieur, S. Multicompartment lipid cubic nanoparticles with high protein upload: Millisecond dynamics of formation. *ACS Nano* **2014**, *8*, 5216–5226. [CrossRef] [PubMed]
36. Angelova, A.; Fajolles, C.; Hocquelet, C.; Djedaini-Pilard, F.; Lesieur, S.; Bonnet, V.; Perly, B.; Lebas, G.; Mauclaire, L. Physico-chemical investigation of asymmetrical peptidolipidyl-cyclodextrins. *J. Colloid Interface Sci.* **2008**, *322*, 304–314. [CrossRef] [PubMed]
37. Chen, Y.Y.; Angelova, A.; Angelov, B.; Drechsler, M.; Garamus, V.M.; Willumeit-Romer, R.; Zou, A.H. Sterically stabilized spongosomes for multidrug delivery of anticancer nanomedicines. *J. Mater. Chem. B* **2015**, *3*, 7734–7744. [CrossRef]
38. Zerkoune, L.; Angelova, A.; Lesieur, S. Nano-assemblies of modified cyclodextrins and their complexes with guest molecules: Incorporation in nanostructured membranes and amphiphile nanoarchitectonics design. *Nanomaterials* **2014**, *4*, 741–765. [CrossRef] [PubMed]
39. Guo, Z.H.; Liu, X.F.; Hu, J.S.; Yang, L.Q.; Chen, Z.P. Synthesis and self-assembled behavior of ph-responsive chiral liquid crystal amphiphilic copolymers based on diosgenyl-functionalized aliphatic polycarbonate. *Nanomaterials* **2017**, *7*, 169. [CrossRef] [PubMed]
40. Collings, P.J.; Hird, M. *Introduction to Liquid Crystals Chemistry and Physics*; Chapter 3; Taylor & Francis: London, UK, 1997.
41. Liu, J.H.; Hung, H.J.; Yang, P.C.; Tien, K.H. Thermal Recordable Novel Cholesteric Liquid Crystalline Polyacrylates Containing Various Chiral Moieties. *J. Polym. Sci. Part A Polym. Chem.* **2008**, *46*, 6214–6228. [CrossRef]
42. Favier, A.; Charreyre, M.T. Experimental requirements for an efficient control of free-radical polymerizations via the reversible addition-fragmentation chain transfer (RAFT) process. *Macromol. Rapid Commun.* **2006**, *27*, 653–692. [CrossRef]
43. Dierking, I. *Textures of Liquid Crystals*; John Wiley & Sons: Hoboken, NJ, USA, 2003; pp. 7–16.

nanomaterials

MDPI

Article

Nanopatterning via Self-Assembly of a Lamellar-Forming Polystyrene-*block*-Poly(dimethylsiloxane) Diblock Copolymer on Topographical Substrates Fabricated by Nanoimprint Lithography

Dipu Borah [1,*], Cian Cummins [1], Sozaraj Rasappa [1], Ramsankar Senthamaraikannan [1], Mathieu Salaun [2], Marc Zelsmann [2], George Liontos [3], Konstantinos Ntetsikas [3], Apostolos Avgeropoulos [3] and Michael A. Morris [1,*]

1 AMBER Centre & CRANN, Trinity College Dublin, College Green, Dublin, Ireland; cumminci@tcd.ie (C.C.); r_sola27@yahoo.co.in (S.R.); ramsankar.s@tcd.ie (R.S.)

2 Laboratoire des Technologies de la Microelectronique (CNRS), 38054 Grenoble, France; mathieu.salaun@cea.fr (M.S.); marc.zelsmann@cea.fr (M.Z.)

3 Department of Materials Science Engineering, University of Ioannina, University Campus-Dourouti, 45110 Ioannina, Greece; gliontos@cc.uoi.gr (G.L.); kntetsik@cc.uoi.gr (K.N.); aavger@cc.uoi.gr (A.A.)

* Correspondence: borahd@tcd.ie (D.B.); morrism2@tcd.ie (M.A.M.); Tel.: +353-1-896-3089 (M.A.M)

Received: 8 December 2017; Accepted: 2 January 2018; Published: 9 January 2018

Abstract: The self-assembly of a lamellar-forming polystyrene-*block*-poly(dimethylsiloxane) (PS-*b*-PDMS) diblock copolymer (DBCP) was studied herein for surface nanopatterning. The DBCP was synthesized by sequential living anionic polymerization of styrene and hexamethylcyclotrisiloxane (D_3). The number average molecular weight (M_n), polydispersity index (M_w/M_n) and PS volume fraction (φ_{ps}) of the DBCP were $M_n{}^{PS}$ = 23.0 kg mol^{-1}, $M_n{}^{PDMS}$ = 15.0 kg mol^{-1}, M_w/M_n = 1.06 and φ_{ps} = 0.6. Thin films of the DBCP were cast and solvent annealed on topographically patterned polyhedral oligomeric silsesquioxane (POSS) substrates. The lamellae repeat distance or pitch (λ_L) and the width of the PDMS features (d_L) are ~35 nm and ~17 nm, respectively, as determined by SEM. The chemistry of the POSS substrates was tuned, and the effects on the self-assembly of the DBCP noted. The PDMS nanopatterns were used as etching mask in order to transfer the DBCP pattern to underlying silicon substrate by a complex plasma etch process yielding sub-15 nm silicon features.

Keywords: directed self-assembly; lamellar diblock copolymer; polyhedral oligomeric silsesquioxane (POSS); nanoimprint lithography; pattern transfer

1. Introduction

To maintain historic improvements in the performance of semiconductor electronics (Moore's Law), the dimensions of critical circuit elements are shrinking towards 10 nm [1]. The most sophisticated semiconductor devices, such as microprocessors and memory chips, are patterned with high-resolution projection UV lithography [2]. However, sub-10 nm feature sizes are challenging and expensive to produce [3]. Alternative "top-down" techniques including thermal [4], e-beam [5,6] and X-ray [7] lithographies are largely unproven for manufacturing. Alternatively, the hierarchical self-assembly of molecular building blocks through molecular recognition and molecule-surface interactions [8] may have the potential for forming surface nanopatterns to create circuit elements. However, it is highly challenging to achieve the required long-range translational order and pattern robustness needed for manufacture with bottom-up approaches [9–12].

Long-range order of self-assembled systems can be achieved by directed self-assembly (DSA) where an external potential "guides" the self-assembling materials [13–17]. One form of guiding structure is well-defined surface topography (graphoepitaxy) and nanoimprint lithography (NIL)

is a technique of simple and cost-effective means for generating topographical patterns [18–20], and can be used to generate substrate topography from a well-defined stamp [21]. Nanoimprinting can guide the self-assembly of a diblock copolymer (DBCP), controlling alignment (to surface direction) and orientation (relative to the surface plane), as was first reported by Li et al. on polystyrene-*b*-poly(methylmethacrylate) (PS-*b*-PMMA) [22]. The polystyrene-*b*-poly(dimethylsiloxane) (PS-*b*-PDMS) is a highly promising system and has been amenable to NIL based DSA [23,24]. PS-*b*-PDMS has a high Flory–Huggins parameter ($\chi = 68/T - 0.037$) allowing formation of small microdomain sizes [25].

Work on the PS-*b*-PDMS system has been very largely limited to cylinder forming systems, and the lamellar forming compositions are expected to have a number of challenges, such as surface dewetting due to high hydrophobicity, and the formation of wetting layers [26]. Low PDMS surface energy and the possibility of forming bonds to surface silanol groups can lead to PDMS formation at the air and substrate interface [27], and for lamellar systems, this is likely to result in parallel orientation and 2D sheet-like structures. To enable proper graphoepitaxy, it is necessary to not only control sidewall chemistry, but also to produce appropriate surface chemistry to define pattern orientation.

In this work, we have attempted to define the required surface chemistry to demonstrate the practicality of using lamellar PS-*b*-PDMS systems. Polyhedral oligomeric silsesquioxane (POSS) materials can be tailored to allow definition of surface chemistry derivatives and three such POSS materials were studied here. It is shown that POSS can facilitate vertical orientation of lamellar. However, the neutral surfaces do not facilitate alignment of the patterns in graphoepitaxial structures, but can be used to understand surface chemistry effects in these systems.

2. Results and Discussion

2.1. DBCP Self-Assembly on PDMS-OH Brush Coated Substrate

The details of the steps involved in the fabrication of POSS template and the subsequent DBCP self-assembly and plasma etching are presented in Scheme 1. The lamellar-forming PS-*b*-PDMS DBCP deposited on PDMS-OH brush (thickness ~4.3 nm as measured by ellipsometry) showed microphase separation after solvent annealing in toluene, as is evident from the SEM image of samples after ETCH1, Figure 1a. Note that no pattern was observed using Atomic Force Microscopy (AFM) studies of unetched, solvent-annealed films, suggesting that a PDMS wetting layer is formed at the air–polymer interface. The PDMS-OH brush favors PDMS (brush)–PDMS (DBCP) nteractions and a significantly lower PDMS surface energy, compared to PS results in a wetting PDMS layer at the substrate–polymer interface and at the polymer–air interface [24,26,27]. Note, however, that the SEM image shown in Figure 1a is not representative of the whole surface, since the DBCP dewets (Figure 1b) during solvent annealing (good coverage is observed after film casting). The films at the brush layers are generally poor, exhibiting multilayer pattern formation, low lamellar correlation lengths and defects such as disclinations, dislocations, etc. The SEM determined lamellae repeat distance or pitch (λ_L), and the PDMS lamellae width (d_L) are ~35 nm and ~17 nm.

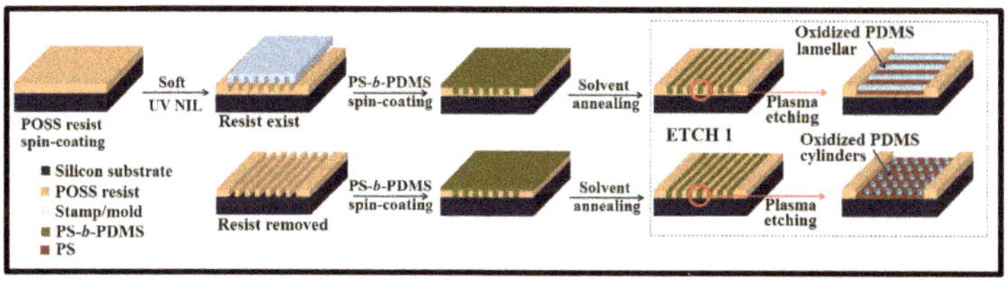

Scheme 1. Schematic depicting the process flows of fabricating nanopatterned POSS template by the soft UV-NIL process, PS-*b*-PDMS self-assembly and the ETCH1 process. Two processes are shown, with and without residual resist removal.

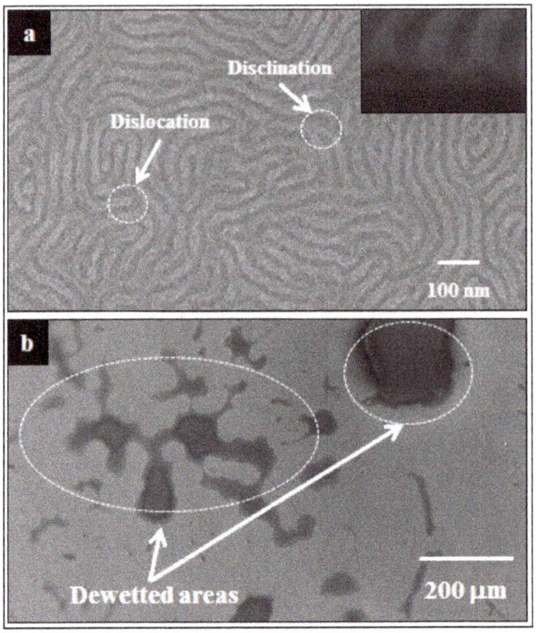

Figure 1. (**a**) Top-down SEM image (inset, cross-section image). Various defects noted in the micrograph and (**b**) top-down SEM images showing the coverage of microphase separated PS-*b*-PDMS film (as revealed by ETCH1 (CF_4 and O_2 etches)).

2.2. *Effect of POSS Type on DBCP Self-Assembly on Planar POSS Coated Substrates*

In an effort to improve the DBCP patterns, POSS materials were developed as an improved alternative to polymer brush. The surface properties of the POSS films can be controlled by grafting different ligands to the POSS cages, as is revealed from the data compiled in Table 1 for POSS-A, POSS-G, and POSS-C6, where the POSS cage is functionalized with eight aliphatic acrylo, glycidyl, and epoxide ligands, respectively. The structural formulae of the POSS derivatives are shown in Scheme 2. As seen in Table 1, POSS-A is the material of highest hydrophilicity, whilst POSS-C6 is the most hydrophobic. It is worthy to make a comparison of the surface free energies of POSSs with that of PS and PDMS. The surface free energies of PS and PDMS are 29.9 and 19.8 mNm^{-1}, respectively [28,29]. It is evident that the surface free energies of POSS derivatives are higher than both PS and PDMS. It is understood that a polymer brush made of the minority block (PDMS here) that preferentially wets the substrate, is likely to have little effect on ordering of the DBCP [30]. Apparently, the POSSs substrates with higher surface free energy likely to provide the perfect energy barrier for the diffusion of PS-*b*-PDMS DBCP into the resist surface.

Table 1. Contact angle (θ) and surface free energy (*SFE*) of POSS-A, POSS-G, and POSS-C6.

POSS Type	θ_{DI} (°)	θ_{DIM} (°)	θ_{EG} (°)	*SFE* (mN/m)
POSS-A	59.3	42.7	33.9	47.5
POSS-G	68.1	43.0	44.9	42.7
POSS-C6	83.6	57.3	65.1	31.1

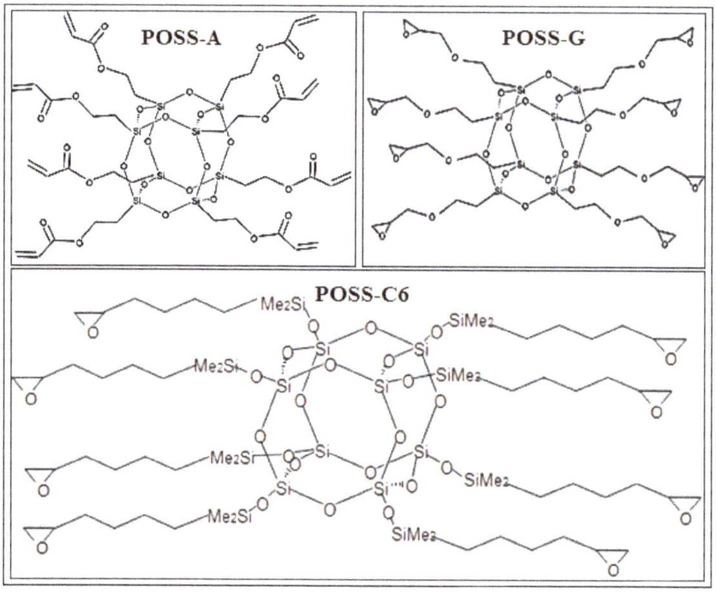

Scheme 2. Structural details of the POSS resists used in the present study.

Self-assembly of the PS-*b*-PDMS DBCP was followed on POSS modified planar substrates, and illustrative data presented in Figure 2. The surface coverage was much higher than on brush modified surfaces, with greater than 80% coverage being obtained in all cases (Figure 2). As well as the coverage advantages, it should be noted that POSS deposition is significantly simpler than brush attachment procedures. Note, however, that POSSs being silicon based compounds, might have a tendency to wet the PDMS block of the DBCP, forming a PDMS wetting layer at the DBCP–POSS interface. However, it is very difficult to isolate the PDMS wetting layer from the POSS resist in the cross-section SEM images (inset, Figure 2), because of the similar contrast. Figure 2 shows that the DBCP forms lamellar patterns for POSS-G and POSS-C6, however, in POSS-A, the morphology is more complex, showing well-defined regions of lamellar patterns and more irregular arrangements of the domains. The less well-defined areas have domains of differing sizes and shapes, ranging from pseudospherical to stripes. There are regions reminiscent of the phase transition between the lamellar phase and cylindrical phase found in the hexagonal perforated lamellar (HPL) structure [31]. Since this polymer composition is 60:40 for PS/PDMS, respectively, it is close to the phase region between the lamellar and cylinder structures where the HPL and IA3 gyroid structures exist, and this may be a complex mixture of these phases. The reason that this complex phase is only seen on the POSS-A film may be attributed to the fact that the surface chemistry favors a PDMS wetting layer, which effectively decreases the content of PDMS in the interior of the DBCP film and moves the composition towards the hexagonal regions.

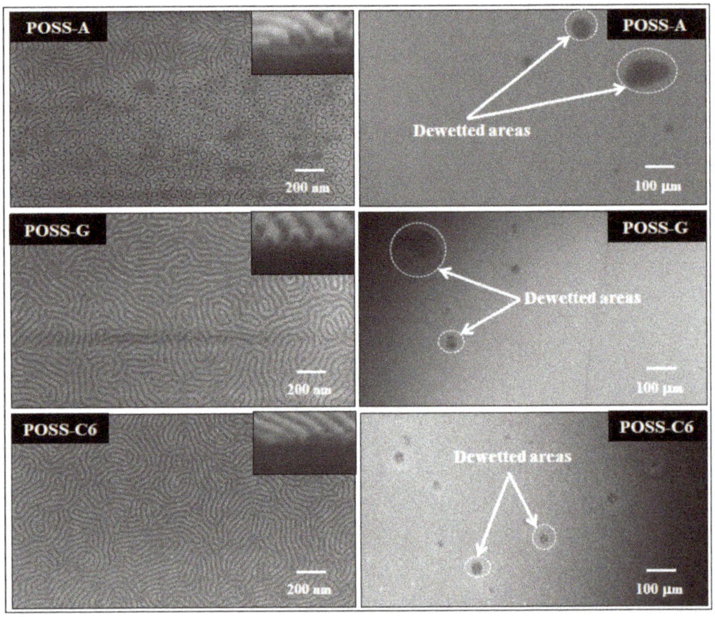

Figure 2. Top-down SEM images of lamellar PS-*b*-PDMS patterns (**left side**) (inset, cross-section images) and coverages of microphase separated PS-*b*-PDMS films (**right side**) following ETCH1 at silicon substrates modified with various POSS thin films.

2.3. *Effect of POSS Topography on DBCP Self-Assembly*

The directed self-assembly (DSA) of PS-*b*-PDMS has been demonstrated for cylinder-forming PS-*b*-PDMS in several articles [26,32–39], however, there are only a few reports of lamellar-forming PS-*b*-PDMS systems [40,41]. In an effort to understand the effect of surface chemistries further, the lamellar-forming PS-*b*-PDMS DBCP films were deposited on NIL patterned POSS substrates. These topographical patterns are important because preferential interactions with blocks can be viewed in top down images. These NIL defined POSS topographies have been discussed elsewhere [24,32,42]. Briefly, NIL produces well defined channel or trench like structures. The channels were 50 nm deep, 270 nm wide, and the channel spacing was 500 nm. Also present is a 15 nm POSS resist layer residue. The deposition of DBCP was controlled using solution concentration and spin speed to just fill the channels so that the thickness is similar to the L_0 of the DBCP, favoring vertical alignment of the domains. It should also be stated that the POSS structures are quite stable during the prolonged solvent anneal procedures when the DBCP is present (Figure 3).

Preferential alignment of the lamellar to the channel side wall was not observed here. However, the topography allowed further understanding of the effects of surface chemistry changes. Figure 3 shows the results of PS-*b*-PDMS self-assembly on POSS topography in the presence and absence of the residual layer. With the resist layer present, all of the POSS structures provided lamellar arrangements. It is clear from the data presented in Figures 2 and 3 that the POSS surfaces are essentially neutral, and do not favor either block. This neutrality is quite clearly seen in Figure 3, because the sidewalls favor an orthogonal arrangement of lamellar (i.e., across the trench), and both PS and PDMS lamellar exist at the sidewall. This is somewhat unexpected for POSS-A, since planar substrates yielded the HPL type arrangement. On this POSS, there is an obvious fine balance in energies that make both the HPL and lamellar structure approximately equally likely. However, the presence of topography makes the lamellar structure slightly more prevalent. When the POSS residual layer is removed, HPL type structures tend to dominate, suggesting the trench base is no longer neutral to both blocks, allowing the formation of a wetting layer and effective compositional changes in the interior of the POSS films. However, the sidewalls remain neutral, with both PS and PDMS present.

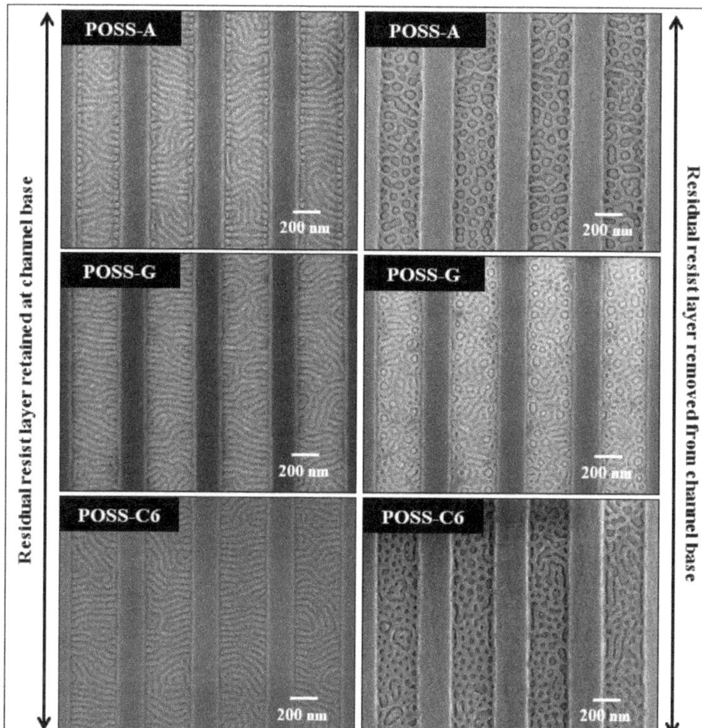

Figure 3. Top-down SEM images of lamellar-forming PS-*b*-PDMS at topographical POSS substrates before and after residual resist layer removal and following ETCH1.

2.4. Pattern Transfer to Underlying Substrate for Nanopatterning

Pattern transfer both demonstrates the regularity of the polymer structure through the film (i.e., there is no complex 3D morphology) but also that the structure has practicality for fabrication of nanodimensioned substrate features. ETCH2 and ETCH3 were used to generate silicon features at both planar and topographical substrates. Figure 4 shows the top-down and cross-section SEM images of the silicon nanorod-type structures on a brush terminated planar silicon substrate and a POSS-C6 patterned silicon substrate. The pattern transfer is successful in both cases, and in all systems studied. Note that on the patterned substrates, the channel mesas have been largely removed, as they are clearly as etch resistant as the oxidized PDMS structures formed by ETCH1. The cross-section SEM image from the planar substrate shows that the silicon nanorod-like structure that is produced via this complex etch procedure has a feature size of ~14 nm, with a feature height of ~38 nm on the planar substrate. The nanoscale silicon features are slightly narrower in width than that of the initial oxidized PDMS features, due to a partly isotropic etch process. The PDMS patterns obtained on POSS-C6 patterned substrate without residual resist layer at the channel base upon pattern transfer, produced silicon nanorod-like features of thickness ~13 nm and a height of ~38 nm.

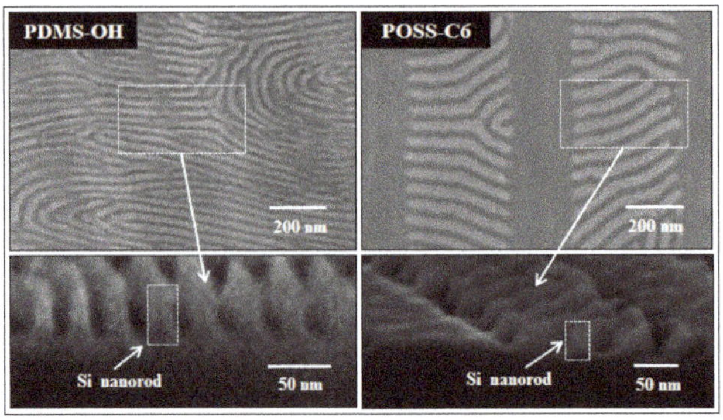

Figure 4. Top-down and high-resolution cross-section SEM images of PDMS patterns transferred to brush modified planar substrates and POSS-C6 patterned silicon substrates.

3. Materials and Methods

3.1. Synthesis of Polymers and Molecular Characteristics

The lamellar-forming PS-*b*-PDMS DBCP was synthesized by sequential living anionic polymerization of styrene and hexamethylcyclotrisiloxane (D_3), as described in the Supplementary Materials. The number average molecular weight (M_n) and polydispersity index (M_w/M_n) of the DBCP were determined using size exclusion chromatography (SEC), and determined to be $M_n{}^{PS}$ = 23.0 kg mol^{-1}, $M_n{}^{PDMS}$ = 15.0 kg mol^{-1}, and M_w/M_n = 1.06. The volume fraction (φ_{ps}) of PS in the DBCP was determined by ^{1}H-nuclear magnetic resonance (NMR) spectroscopy at 0.60. The hydroxyl-terminated PDMS (PDMS-OH) polymer brush was also synthesized via anionic polymerization techniques, and end-capped with 1–2 monomeric units of ethylene oxide (EO) in pyridine, and terminated with methanol (MeOH). M_n and M_w/M_n were 5.5 kg mol^{-1} and 1.06, respectively. Full details of molecular characterization are available in Supporting Information.

3.2. POSS Materials and Synthesis of POSS-C6

OctaSilane POSS® (POSS), Acrylo POSS® (POSS-A) and Glycidyl POSS® (POSS-G) were purchased from Hybrid Plastics (Hattiesburg, MS, USA). 1,2-epoxy-5-hexene, styrene, platinum(0)-1,3-divinyl-1,1,3,3-tetramethyldisiloxane complex solution (Karstedt´s catalyst), toluene and ethyl-L-lactate were purchased from Sigma Aldrich (Dublin, Ireland). The photoinitiators Irgacure® 250 (iodonium(4-methylphenyl)[4-(2-methylpropyl)phenyl]-hexafluorophosphonate, 75 wt % in propylene carbonate), Irgacure® 819 (bis(2,4,6-trimethylbenzoyl)-phenylphosphineoxide) and the sensitizer Genocure® ITX (isopropyl thioxanthone) were generously provided by, respectively, BASF Resins (Berlin, Germany) and RAHN AG Energy Curing (Zurich, Switzerland). The epoxy-functionalized POSS monomer POSS-C6 was prepared according to methodology detailed previously [43]. OctaSilane POSS cages bearing eight dimethylsilyloxy groups were functionalized by a hydrosilylation reaction carried out on the Si–H functions. Typically, 2 g (1.96 mmol) of OctaSilane POSS were dissolved in 10 mL of anhydrous toluene. Subsequently, 1.96*8 mmol (8 ligands) of 1,2-epoxy-5-hexene (+10% excess) were added to the mixture with 2 drops of Karstedt's catalyst. The reaction was carried out at 353 K during 3 h under argon. Solvent, catalyst, and material were removed under vacuum. Characterization details are available in Supporting Information.

3.3. Resist Preparation and Fabrication of POSS Templates by UV-NIL

Topographical substrates for graphoepitaxy experiments were fabricated using UV nanoimprint lithography and a PDMS elastomeric mold (so-called "soft UV-NIL"). POSS monomers were diluted in propylene glycol methyl ether acetate (PGMEA). Two moles (relative to epoxy groups) of Irgacure® 250 photo-initiator and 0.5 wt % of Genocure® ITX photo sensitizer were added to provide UV sensitivity. These solutions were spin-coated onto 4" silicon (B doped, *p*-type, thickness 650 μm, and resistivity 6–14 ohm-cm) wafers of <100> orientation and a native oxide layer of ~2 nm, to form 50 nm thick resist films. Some of the POSS coated substrates were not patterned for planar surface studies. The PDMS mold (forming 50 nm deep, 270 nm wide trenches in the resist) was stamped into the resist layer at a pressure of 200 kPa, and the stack was then exposed to UV radiation (365 nm wavelength) for 3 min. Prior to imprinting, molds were treated with a fluorosilane anti-sticking compound (Optool DSX, Daikin Chemical, Dusseldorf, Germany) to allow even demolding. The resist residual layer at the bottom of the imprinted trenches (approx. 15 nm thick) could be removed by a CF_4 (15 sccm) inductively coupled plasma (ICP) etch 4–8 s (varies with the resist type), which has been detailed before [21].

3.4. Deposition of PDMS-OH Brush on Silicon Substrates

Silicon substrates were ultrasonically degreased in acetone and IPA solutions for 5 min each, dried in flowing N_2 gas, and baked for 2 min at 393 K in an ambient atmosphere, to remove any residual IPA. They were then cleaned in piranha solution (1:3 v/v 30% H_2O_2/H_2SO_4) at 363 K for 60 min, rinsed with deionized water, acetone, and ethanol before drying under N_2 flow. Piranha activation removes contamination, and generates substrate surface hydroxyl groups. Hydroxyl-terminated PDMS brush solutions (1.0 wt % in toluene) were spin-coated (P6700 Series spin-coater, Specialty Coating Systems, Indianapolis, IN, USA) onto the wafers at 3000 rpm for 30 s. Samples were annealed in a vacuum oven (Townson & Mercer EV018, Wolflabs, York, UK) at 443 K under vacuum (1.3 kPa of residual pressure) for 6 h. This allows condensation of the brush hydroxyl end groups with surface silanol groups, resulting in polymer chain brushes being chemically attached to the substrate. Unbound polymers were removed by sonication (Cole-Palmer 8891 sonicator) and rinsing in toluene.

3.5. Deposition of PS-b-PDMS and Solvent Annealing

Thin films of PS-*b*-PDMS were prepared by depositing dilute solutions (e.g., 1.0 wt %) of the DBCP in toluene onto polymer brush anchored and unpatterned/patterned POSS substrates by spin coating (e.g., 3200 rpm and 30 s). As-cast thin films were solvent annealed in glass jars under saturated toluene environment at room temperature (~288 K) for 3 h. After removal solvent was allowed to evaporate at ambient conditions.

3.6. Plasma Etching of PS-b-PDMS Films and Pattern Transfer

All samples were plasma etched to allow easy imaging of the DBCP patterns and to transfer the polymer pattern to the substrate. Details of the etch processes are given elsewhere [33,34]. Briefly, an etch process (ETCH1) was used to convert the PS-*b*-PDMS pattern onto a silicon oxide-like pattern. A CF_4 was used to remove any PDMS wetting layer at the air–polymer interface. An O_2 (30 sccm) plasma etch was used to oxidize the PDMS into a silica-like structure. The steps used follow a similar methodology to that developed by Ross et al. [26]. The process removes the PS component and forms an oxidized form of PDMS on the substrate. These oxidized PDMS patterns were then used as an etch mask for pattern transfer (i.e., ETCH2). This etch consists of a CHF_3/Ar plasma, and removes any non-oxidized PDMS from the surface. This process was followed by a selective silicon etch using CHF_3 and SF_6 to transfer the patterns into the underlying substrate. Any remaining polymer derived material was removed by ETCH3. The residual oxidized PDMS features were removed by a 10 s silica etch based on CHF_3 and Ar. This was followed by an O_2 etch to remove any residual PS and polymer brush. All etching was carried out in an OIPT Plasmalab System100 ICP180 (Oxford Instruments, Bristol, UK) etch tool.

3.7. Material Characterization

Contact angles and surface free energy measurements were carried out using a Krüss DSA 100 (Krüss Optronic, Hamburg, Germany) goniometer. Contact angles were measured by the static sessile drop method and surface free energy was calculated from the measured contact angles of deionized water (DI), diiodomethane (DIM), and ethylene glycol (EG) using the Owens–Wendt model [44]. Film thickness (an average of five readings from different sample areas) was determined by ellipsometry (Plasmos SD2000 Ellipsometer, Filmetrics, Surrey, UK) at an incidence angle of 70°. A Varian IR660 (Agilent Technologies, Cheshire, UK) infrared spectrometer was used to record FTIR data. The measurements were performed in the spectral range of 4000–500 cm^{-1}, with a resolution of 4 cm^{-1} and data averaged over 32 scans. Surface morphology and silicon nanostructures were investigated by scanning electron microscope (SEM) images, and were obtained by a high resolution (<1 nm) Field Emission Zeiss Ultra Plus-SEM (Carl Zeiss AG, Oberkochen, Germany) with a Gemini® column operating at an accelerating voltage of 5 kV. The profile images of the surfaces were also used to measure the film thicknesses which agreed well with ellipsometer data.

4. Conclusions

We have demonstrated in this work the self-assembly and directed self-assembly (DSA) of a lamellar-forming PS-*b*-PDMS diblock copolymer (DBCP) system on substrates coated with PDMS-OH brush and POSS materials. The behavior of POSS materials in terms of wetting the DBCP domains can be obtained from DSA of the DBCP on patterned POSS substrates fabricated by soft UV nanoimprint lithography. The functionalization of the POSS materials have shown advantages of tuning the surface properties to improve the wetting property of the DBCP, direct self-assembly and pattern orientation/alignment. The lamellar DBCP forms oxidized PDMS patterns that can be successfully pattern transferred to underlying silicon to fabricate silicon nanoscale features with sub-15 nm feature size.

Supplementary Materials: The following are available online at http://www.mdpi.com/2079-4991/8/1/32/s1, Details of the synthesis and characterization of PS-*b*-PDMS DBCP, PDMS-OH polymer brush and characterization of synthesized POSS-C6.

Acknowledgments: Financial support for this work is provided by the EU FP7 NMP project, LAMAND (grant number 245565) project and the Science Foundation Ireland (grant number 09/IN.1/602), and gratefully acknowledged.

Author Contributions: Dipu Borah and Michael A. Morris conceived and designed the experiments; Dipu Borah, Cian Cummins, Sozaraj Rasappa and Ramsankar Senthamaraikannan performed the DBCP experiments and characterization; Mathieu Salaun and Marc Zelsman performed the NIL experiments; George Liontos, Konstantinos Ntetsikas and Apostolos Avgeropoulos synthesized the DBCP. All authors analysed and discussed the data. Dipu Borah and Michael A. Morris wrote the manuscript.

Conflicts of Interest: The authors declare no conflict of interest.

References

1. Morris, M.A. Directed self-assembly of block copolymers for nanocircuitry fabrication. *Microelectron. Eng.* **2015**, *132*, 207–217. [CrossRef]
2. Wissen, M.; Bogdanski, N.; Moellenbeck, S.; Scheer, H.C. Strategies for hybrid techniques of UV lithography and thermal nanoimprint. In Proceedings of the 2008 24th European Mask and Lithography Conference (EMLC), Dresden, Germany, 21–24 January 2008; pp. 1–11.
3. Xia, Y.; Whitesides, G.M. Soft lithography. *Annu. Rev. Mater. Sci.* **1998**, *28*, 153–184. [CrossRef]
4. Chung, S.; Felts, J.R.; Wang, D.; King, W.P.; Yoreo, J.J.D. Temperature-dependence of ink transport during thermal dip-pen nanolithography. *Appl. Phys. Lett.* **2011**, *99*, 193101. [CrossRef]
5. Grigorescu, A.E.; Hagen, C.W. Resists for sub-20-nm electron beam lithography with a focus on HSQ: State of the art. *Nanotechnology* **2009**, *20*, 292001. [CrossRef] [PubMed]
6. Namatsu, H.; Watanabe, Y.; Yamazaki, K.; Yamaguchi, T.; Nagase, M.; Ono, Y.; Fujiwara, A.; Horiguchi, S. Influence of oxidation temperature on Si-single electron transistor characteristics. *J. Vac. Sci. Technol. B* **2003**, *21*, 2869–2873. [CrossRef]

7. Hirai, Y.; Hafizovic, S.; Matsuzuka, N.; Korvink, J.G.; Tabata, O. Validation of X-ray lithography and development simulation system for moving mask deep X-ray lithography. *J. Microelectromech. Syst.* **2006**, *15*, 159–168. [CrossRef]
8. Hamley, I.W. Nanotechnology with Soft Materials. *Angew. Chem. Int. Ed.* **2003**, *42*, 1692–1712. [CrossRef] [PubMed]
9. Kumar, P. Directed Self-Assembly: Expectations and Achievements. *Nanoscale Res. Lett.* **2010**, *5*, 1367–1376. [CrossRef] [PubMed]
10. Katsuhiko, A.; Jonathan, P.H.; Michael, V.L.; Ajayan, V.; Richard, C.; Somobrata, A. Challenges and breakthroughs in recent research on self-assembly. *Sci. Technol. Adv. Mater.* **2008**, *9*, 014109.
11. Hawker, C.J.; Russell, T.P. Block copolymer lithography: Merging "bottom-up" with "top-down" processes. *MRS Bull.* **2005**, *30*, 952–966. [CrossRef]
12. Michele Perego, G.S. Self-assembly strategies for the synthesis of functional nanostructured materials. *Nuovo Cimento Riv. Ser.* **2016**, *39*, 279–312.
13. Bang, J.; Jeong, U.; Ryu, D.Y.; Russell, T.P.; Hawker, C.J. Block copolymer nanolithography: Translation of molecular level control to nanoscale patterns. *Adv. Mater.* **2009**, *21*, 4769–4792. [CrossRef] [PubMed]
14. Kim, B.H.; Kim, J.Y.; Kim, S.O. Directed self-assembly of block copolymers for universal nanopatterning. *Soft Matter* **2013**, *9*, 2780–2786. [CrossRef]
15. Kim, E.; Ahn, H.; Park, S.; Lee, H.; Lee, M.; Lee, S.; Kim, T.; Kwak, E.-A.; Lee, J.H.; Lei, X.; et al. Directed Assembly of High Molecular Weight Block Copolymers: Highly Ordered Line Patterns of Perpendicularly Oriented Lamellae with Large Periods. *ACS Nano* **2013**, *7*, 1952–1960. [CrossRef] [PubMed]
16. Choi, E.; Park, S.; Ahn, H.; Lee, M.; Bang, J.; Lee, B.; Ryu, D.Y. Substrate-Independent Lamellar Orientation in High-Molecular-Weight Polystyrene-*b*-poly(methyl methacrylate) Films: Neutral Solvent Vapor and Thermal Annealing Effect. *Macromolecules* **2014**, *47*, 3969–3977. [CrossRef]
17. Cummins, C.; Ghoshal, T.; Holmes, J.D.; Morris, M.A. Strategies for Inorganic Incorporation using Neat Block Copolymer Thin Films for Etch Mask Function and Nanotechnological Application. *Adv. Mater.* **2016**, *28*, 5586–5618. [CrossRef] [PubMed]
18. Guo, L.J. Nanoimprint Lithography: Methods and Material Requirements. *Adv. Mater.* **2007**, *19*, 495–513. [CrossRef]
19. Gates, B.D.; Xu, Q.; Stewart, M.; Ryan, D.; Willson, C.G.; Whitesides, G.M. New Approaches to Nanofabrication: Molding, Printing, and Other Techniques. *Chem. Rev.* **2005**, *105*, 1171–1196. [CrossRef] [PubMed]
20. Mårtensson, T.; Carlberg, P.; Borgström, M.; Montelius, L.; Seifert, W.; Samuelson, L. Nanowire Arrays Defined by Nanoimprint Lithography. *Nano Lett.* **2004**, *4*, 699–702. [CrossRef]
21. Austin, M.D.; Ge, H.; Wu, W.; Li, M.; Yu, Z.; Wasserman, D.; Lyon, S.A.; Chou, S.Y. Fabrication of 5 nm linewidth and 14 nm pitch features by nanoimprint lithography. *Appl. Phys. Lett.* **2004**, *84*, 5299–5301. [CrossRef]
22. Li, H.-W.; Huck, W.T.S. Ordered Block-Copolymer Assembly Using Nanoimprint Lithography. *Nano Lett.* **2004**, *4*, 1633–1636. [CrossRef]
23. Park, S.-M.; Liang, X.; Harteneck, B.D.; Pick, T.E.; Hiroshiba, N.; Wu, Y.; Helms, B.A.; Olynick, D.L. Sub-10 nm Nanofabrication via Nanoimprint Directed Self-Assembly of Block Copolymers. *ACS Nano* **2011**, *5*, 8523–8531. [CrossRef] [PubMed]
24. Salaun, M.; Zelsmann, M.; Archambault, S.; Borah, D.; Kehagias, N.; Simao, C.; Lorret, O.; Shaw, M.T.; Sotomayor Torres, C.M.; Morris, M.A. Fabrication of highly ordered sub-20 nm silicon nanopillars by block copolymer lithography combined with resist design. *J. Mater. Chem. C* **2013**, *1*, 3544–3550. [CrossRef]
25. Politakos, N.; Ntoukas, E.; Avgeropoulos, A.; Krikorian, V.; Pate, B.D.; Thomas, E.L.; Hill, R.M. Strongly segregated cubic microdomain morphology consistent with the double gyroid phase in high molecular weight diblock copolymers of polystyrene and poly(dimethylsiloxane). *J. Polym. Sci. Pol. Phys.* **2009**, *47*, 2419–2427. [CrossRef]
26. Jung, Y.S.; Ross, C.A. Orientation-Controlled Self-Assembled Nanolithography Using a Polystyrene–Polydimethylsiloxane Block Copolymer. *Nano Lett.* **2007**, *7*, 2046–2050. [CrossRef] [PubMed]
27. O'Driscoll, B.M.D.; Kelly, R.A.; Shaw, M.; Mokarian-Tabari, P.; Liontos, G.; Ntetsikas, K.; Avgeropoulos, A.; Petkov, N.; Morris, M.A. Achieving structural control with thin polystyrene-*b*-polydimethylsiloxane block copolymer films: The complex relationship of interface chemistry, annealing methodology and process conditions. *Eur. Polym. J.* **2013**, *49*, 3445–3454. [CrossRef]

28. Winesett, D.A.; Story, S.; Luning, J.; Ade, H. Tuning Substrate Surface Energies for Blends of Polystyrene and Poly(methyl methacrylate). *Langmuir* **2003**, *19*, 8526–8535. [CrossRef]
29. Bracic, M.; Mohan, T.; Kargl, R.; Griesser, T.; Hribernik, S.; Kostler, S.; Stana-Kleinschek, K.; Fras-Zemljic, L. Preparation of PDMS ultrathin films and patterned surface modification with cellulose. *RSC Adv.* **2014**, *4*, 11955–11961. [CrossRef]
30. Harrison, C.; Chaikin, P.M.; Huse, D.A.; Register, R.A.; Adamson, D.H.; Daniel, A.; Huang, E.; Mansky, P.; Russell, T.P.; Hawker, C.J.; et al. Reducing Substrate Pinning of Block Copolymer Microdomains with a Buffer Layer of Polymer Brushes. *Macromolecules* **2000**, *33*, 857–865. [CrossRef]
31. Ly, D.Q.; Honda, T.; Kawakatsu, T.; Zvelindovsky, A.V. Kinetic Pathway of Gyroid-to-Cylinder Transition in Diblock Copolymer Melt under an Electric Field. *Macromolecules* **2007**, *40*, 2928–2935. [CrossRef]
32. Simao, C.; Francone, A.; Borah, D.; Lorret, O.; Salaun, M.; Kosmala, B.; Shaw, M.T.; Dittert, B.; Kehagias, N.; Zelsmann, M.; et al. Soft Graphoepitaxy of Hexagonal PS-*b*-PDMS on Nanopatterned POSS Surfaces fabricated by Nanoimprint Lithography. *J. Photopolym. Sci. Technol.* **2012**, *25*, 239–244. [CrossRef]
33. Borah, D.; Senthamaraikannan, R.; Rasappa, S.; Kosmala, B.; Holmes, J.D.; Morris, M.A. Swift Nanopattern Formation of PS-*b*-PMMA and PS-*b*-PDMS Block Copolymer Films Using a Microwave Assisted Technique. *ACS Nano* **2013**, *7*, 6583–6596. [CrossRef] [PubMed]
34. Borah, D.; Shaw, M.T.; Holmes, J.D.; Morris, M.A. Sub-10 nm Feature Size PS-*b*-PDMS Block Copolymer Structures Fabricated by a Microwave-Assisted Solvothermal Process. *ACS Appl. Mater. Interfaces* **2013**, *5*, 2004–2012. [CrossRef] [PubMed]
35. Borah, D.; Simao, C.D.; Senthamaraikannan, R.; Rasappa, S.; Francone, A.; Lorret, O.; Salaun, M.; Kosmala, B.; Kehagias, N.; Zelsmann, M.; et al. Soft-graphoepitaxy using nanoimprinted polyhedral oligomeric silsesquioxane substrates for the directed self-assembly of PS-*b*-PDMS. *Eur. Polym. J.* **2013**, *49*, 3512–3521. [CrossRef]
36. Borah, D.; Rasappa, S.; Senthamaraikannan, R.; Kosmala, B.; Shaw, M.T.; Holmes, J.D.; Morris, M.A. Orientation and Alignment Control of Microphase-Separated PS-*b*-PDMS Substrate Patterns via Polymer Brush Chemistry. *ACS Appl. Mater. Interfaces* **2012**, *5*, 88–97. [CrossRef] [PubMed]
37. Borah, D.; Ozmen, M.; Rasappa, S.; Shaw, M.T.; Holmes, J.D.; Morris, M.A. Molecularly Functionalized Silicon Substrates for Orientation Control of the Microphase Separation of PS-*b*-PMMA and PS-*b*-PDMS Block Copolymer Systems. *Langmuir* **2013**, *29*, 2809–2820. [CrossRef] [PubMed]
38. Hobbs, R.G.; Farrell, R.A.; Bolger, C.T.; Kelly, R.A.; Morris, M.A.; Petkov, N.; Holmes, J.D. Selective Sidewall Wetting of Polymer Blocks in Hydrogen Silsesquioxane Directed Self-Assembly of PS-*b*-PDMS. *ACS Appl. Mater. Interfaces* **2012**, *4*, 4637–4642. [CrossRef] [PubMed]
39. Jung, Y. S; Jung, W.; Tuller, H.L.; Ross, C.A. Nanowire Conductive Polymer Gas Sensor Patterned Using Self-Assembled Block Copolymer Lithography. *Nano Lett.* **2008**, *8*, 3776–3780. [CrossRef] [PubMed]
40. Son, J.G.; Gotrik, K.W.; Ross, C.A. High-Aspect-Ratio Perpendicular Orientation of PS-*b*-PDMS Thin Films under Solvent Annealing. *ACS Macro Lett.* **2012**, *1*, 1279–1284. [CrossRef]
41. Bai, W.; Gadelrab, K.; Alexander-Katz, A.; Ross, C.A. Perpendicular Block Copolymer Microdomains in High Aspect Ratio Templates. *Nano Lett.* **2015**, *15*, 6901–6908. [CrossRef] [PubMed]
42. Borah, D.; Rasappa, S.; Salaun, M.; Zellsman, M.; Lorret, O.; Liontos, G.; Ntetsikas, K.; Avgeropoulos, A.; Morris, M.A. Soft Graphoepitaxy for Large Area Directed Self-Assembly of Polystyrene-block-Poly(dimethylsiloxane) Block Copolymer on Nanopatterned POSS Substrates Fabricated by Nanoimprint Lithography. *Adv. Funct. Mater.* **2015**, *25*, 3425–3432. [CrossRef]
43. Crivello, J.V.; Malik, R. Synthesis and photoinitiated cationic polymerization of monomers with the silsesquioxane core. *J. Polym. Sci. Pol. Chem.* **1997**, *35*, 407–425. [CrossRef]
44. Żenkiewicz, M. Comparative study on the surface free energy of a solid calculated by different methods. *Polym. Test.* **2007**, *26*, 14–19. [CrossRef]

nanomaterials

MDPI

Article

Anomalous Elastic Properties of Attraction-Dominated DNA Self-Assembled 2D Films and the Resultant Dynamic Biodetection Signals of Microbeam Sensors

Junzheng Wu [1], Ying Zhang [1] and Nenghui Zhang [1,2,*]

1 Shanghai Key Laboratory of Mechanics in Energy Engineering, Shanghai Institute of Applied Mathematics and Mechanics, Shanghai University, Shanghai 200072, China; wujunzheng123@126.com (J.W.); zhangxiaoyi@shu.edu.cn (Y.Z.)

2 Department of Mechanics, College of Sciences, Shanghai University, Shanghai 200444, China

* Correspondence: nhzhang@shu.edu.cn

Received: 4 March 2019; Accepted: 20 March 2019; Published: 3 April 2019

Abstract: The condensation of DNA helices has been regularly found in cell nucleus, bacterial nucleoids, and viral capsids, and during its relevant biodetections the attractive interactions between DNA helices could not be neglected. In this letter, we theoretically characterize the elastic properties of double-stranded DNA (dsDNA) self-assembled 2D films and their multiscale correlations with the dynamic detection signals of DNA-microbeams. The comparison of attraction- and repulsion-dominated DNA films shows that the competition between attractive and repulsive micro-interactions endows dsDNA films in multivalent salt solutions with anomalous elastic properties such as tensile surface stresses and negative moduli; the occurrence of the tensile surface stress for the attraction-dominated DNA self-assembled film reveals the possible physical mechanism of the condensation found in organism. Furthermore, dynamic analyses of a hinged–hinged DNA-microbeam reveal non-monotonous frequency shifts due to attraction- or repulsion-dominated dsDNA adsorptions and dynamic instability occurrence during the detections of repulsion-dominated DNA films. This dynamic instability implies the existence of a sensitive interval of material parameters in which DNA adsorptions will induce a drastic natural frequency shift or a jump of vibration mode even with a tiny variation of the detection conditions. These new insights might provide us some potential guidance to achieve an ultra-highly sensitive biodetection method in the future.

Keywords: DNA film; micromechanical biosensor; elastic property; natural frequency; multiscale method

1. Introduction

Unlike the wormlike genomic DNA in dilute solutions, DNA condensation has been regularly found in cell nucleus, bacterial nucleoids, and viral capsids [1–3]. In the condensed state, despite the strong electrostatic repulsion that exists between negatively charged molecules, DNA double helices are locally aligned and separated by just one or two layers of water [1,4], and this indicates the emergence of the attractive interactions induced by multivalent cations, lipids, or polymers [1,3]. Several theoretical works such as attractive electrostatic forces, screened Debye–Hückel interactions, and water-structuring or hydration forces, have tried to explain the physical origin of the attractive interactions [4]. However, the lack of experimental measurements prevented further development and discrimination among these alternative theories [4]. Recently, by the single-molecule experiments using biochemical, osmotic stress, X-ray scattering, optical techniques, and silicon nanotweezers integrating with a microfluidic device, the three-dimensional condensation of DNA in solution has been studied [5,6]. Also, Langevin dynamics simulations have been used to study the DNA condensation in single-molecule experiments [2]. Furthermore, experiments have shown that the structures of Mg cation with deep-ion-binding sites and phosphoester sites make it capable of bridging features, not only along the helix, but also across helix binding [7].

Surface-effect-based nanomechanical biosensor is a unique tool for measuring biomolecular interactions and molecular conformational changes without molecular labeling [8–10]. For instance, as a promotion of the general observation of three-dimensional aggregation of DNA in solution, Mertens et al. provided an alternative method to obtain the direct information about the forces involved in a two-dimensional condensation of DNA by using functionalized DNA- microcantilever sensors [5]. Experiment results give direct evidence that trivalent ions turn the repulsive electrostatic forces between short strands of single-stranded DNA into attractive as a previous step to condensation [5]. Other works also show that different kinds of buffer salt solutions [5], salt concentrations [11], DNA packing densities [11], and environment temperatures [12] will trigger the change of surface stress and the resultant transition of bending direction. Eom et al. revealed that the resonant frequency shift for a microcantilever resonator due to biomolecular adsorption depends on, not only the mass of adsorbed biomolecules, but also the biomolecular interactions [13]. Lee et al. observed an anomalous increase in the resonant frequency during the Au adsorption on the microcantilever, and speculated that the positive frequency shift was ascribed to the variation in the spring constant related to the surface stress [14]. Tamayo et al. also showed that the adsorption position and the thickness ratio between the adsorbed layer and the microbeam induced an anomalous resonant frequency shift [15]. However, the quantified assessment description of the relationship between the anomalous signals and the experiment conditions, especially for the attraction cases, still remains an open question.

Different from the previous analysis of piezoelectric properties of double-stranded DNA (dsDNA) films and its effect on the static detection signals of microcantilevers [16], this paper is devoted to the establishment of a multiscale model to characterize the macroscale elastic properties of dsDNA films and their correlations with the anomalous dynamic detection signals of hinged–hinged microbeams induced by micro-interactions. First, two mesoscopic potentials of free energy for a repulsion-dominated dsDNA film in NaCl solution or attraction-dominated dsDNA films in multivalent salt solutions are used to predict their elastic properties, including surface stress and elastic modulus. The comparative study of attraction- and repulsion-dominated DNA films shows that the competition between attractive and repulsive micro-interactions endows the attraction-dominated dsDNA films with anomalous elastic properties such as tensile surface stress and negative modulus, and the predicted tensile surface stress reveals the possible physical mechanism of the condensation found in organism. Next, the first-order natural frequency shifts of a hinged–hinged microbeam with a repulsion- or attraction-dominated DNA film are discussed. Numerical results show a non-monotonic variation in frequency shifts due to dsDNA adsorptions and totally different responses between detections of attraction-dominated films and that of repulsion-dominated films, and the dynamic instability occurs during the detections of repulsion-dominated DNA films. This instability indicates that there is a sensitive interval of material parameters in which DNA adsorptions will induce a drastic natural frequency shift or a jump of vibration mode from stability to instability even with a tiny variation of the detection conditions. At last, the physical mechanism underlying these non-monotonous variations in detection signals of dsDNA films at different experiment conditions is discussed.

2. Multiscale Analytical Model

In this paper, through the energy method, we are looking forward to establishing a multiscale analytical model to describe the relationship between the surface elastic properties of adsorbed DNA films and the detection signals of DNA-microbeam systems.

Figure 1a shows the scheme of the Atomic Force Microscope (AFM) measurement for biodetections [17], in which a laser is used to capture the adsorption induced deflection of the microcantilever and its reflection is collected by a quadrant photodetector or by a position sensitive detector (PSD). The structure and the relevant coordinate of the microbeam are shown in

Figure 1b. We will investigate a symmetric adsorption with advantages of minimizing both the effects of thermal drift and non-specific binding interactions with the backside of the microcantilever [18,19]. The structure consists of three layers: The two symmetric adsorbed DNA films and the SiN_x/Si substrate with the length of l and the width of b. And E_p and E_s, and h_p and h_s represent the elastic moduli and thicknesses of the self- or directed-assembled DNA film and substrate, respectively. The x-axis is established at the geometric midplane of the substrate, and the positive direction of the z-axis points to the bottom film.

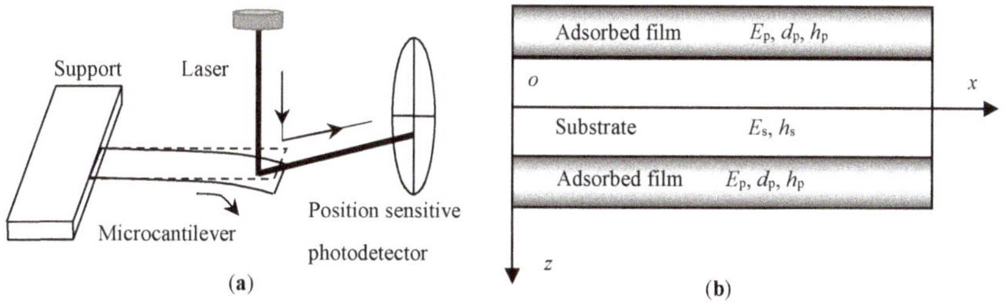

Figure 1. (**a**) Scheme of the Atomic Force Microscope (AFM) measurement; (**b**) schematic showing a microbeam and its coordinate system.

2.1. Elastic Properties of Adsorbed DNA Films

In this section, the adsorbed DNA film is treated as an elastomer. According to continuum mechanics, if the free energy of the self- or directed-assembled DNA film is derived, its elastic properties in a uniaxial compressive/tensile state can be easily obtained as [20]

$$E_p = 3\eta\partial^2 W_b/\partial\varepsilon^2\Big|_{\varepsilon=0}, \ \sigma_p = 3\eta\partial W_b/\partial\varepsilon\big|_{\varepsilon=0} \tag{1}$$

where E_p is the elastic modulus, σ_p is the surface stress, ε is the axial strain, η is the DNA packing density, and $\eta = 2/(\sqrt{3}d_0^2)$ for the hexagonal packing pattern, in which d_0 is the initial interaxial distance [21,22]; W_b is the free energy per unit length between two parallel DNA cylinders. However, there is no a unified formula for the free energy of DNA solutions. In the following section, two mesoscopic free energy potentials will be, respectively, introduced for a repulsion-dominated dsDNA film in NaCl solution or attraction-dominated dsDNA films in multivalent salt solutions.

As for the mesoscopic free energy of dsDNA in multivalent salt solutions, such as spermine $[H_2N(CH_2)_3NH(CH_2)_4NH(CH_2)_3NH_2]$ (valence +4), $Co(NH_3)_6Cl_3$ (valence +3) and sp^{6+} $[H_2N(CH_2)_3NH(CH_2)_3NH(CH_2)_3NH(CH_2)_3NH(CH_2)_3\ NH_2]$ (valence +6), by combining the single-molecule magnetic tweezers and osmotic stress experiments, Todd et al. separated the attractive and repulsive components from the total intermolecular interactions, and proposed an alternative interaction potential of free energy [4]. The free energy per length is given as

$$W_{b1} = \Delta G_{rep} + \Delta G_{att} = \sqrt{3}\lambda(d+\lambda/2)C_R e^{-2d/\lambda}/2 - \sqrt{3}\lambda(d+\lambda)C_A e^{-d/\lambda}, \tag{2}$$

where ΔG_{rep} and ΔG_{att} represent the repulsive and attractive interaction potentials, respectively. By convention, the repulsive interaction potential is defined as positive, and the attractive potential is negative. λ = 4.6 Å is the decay length, C_R and C_A are the corresponding prefactors related to the specific salt conditions, and d is the interaxial distance. According to our previous models [21], the interaxial distance, d, between parallel DNA cylinders after microbeam bending, is given as $d = (1+\varepsilon)d_0$, in which d_0 is the initial interaxial distance, and ε is the bending strain. The thickness of the adsorbed film is approximately taken as the contour length of DNA chain [21,23], namely,

$h_p \approx Na$, N is the DNA nucleotide number, a is the nucleotide length directly obtained from STM experiment, and a = 0.34 nm for dsDNA [24].

As for the mesoscopic free energy of dsDNA in NaCl solution, based on a liquid-crystal model and osmotic pressure experiments, Strey et al. [25] presented a repulsion-dominated interaction potential which has been used to effectively predict the deflection and surface stress of DNA-microbeam systems. The repulsive interaction energy per unit length between two parallel DNA cylinders is given as

$$W_{b2} = W_e + W_h + W_c, \tag{3}$$

where W_e, W_h, and W_c, are, respectively, electrostatic energy, hydration energy, and configurational entropy, and

$$W_e(z,d) = a_0 \exp(-d/\lambda_D)/\sqrt{d/\lambda_D},\ W_h = b_0 \exp(-d/\lambda_H)/\sqrt{d/\lambda_H},$$

$$W_c = c_0 k_B T k_c^{-1/4} \sqrt[4]{\partial^2(W_e+W_h)/\partial d^2 - (1/d)\partial(W_e+W_h)/\partial d},$$

where λ_D is the Debye screening length, λ_H is the correlation length of water [25], a_0, b_0, and c_0 are the fitting parameters for DNA interactions; k_B is the Boltzmann constant, T is the temperature, $k_c = k_B T l_p^{ds}$ is the bending stiffness of a single-molecule dsDNA chain, l_p^{ds} is the persistence length of dsDNA, l_p^{ds}= (50 + 0.0324/I) nm, and I is the buffer salt concentration [26].

Finally, substituting W_{b1} in Equation (2) or W_{b2} in Equation (3) into Equation (1) yields the elastic modulus and surface stress of the adsorbed dsDNA film in multivalent or monovalent NaCl solutions.

2.2. Natural Frequency of DNA-Microbeam System

This section is dedicated to investigating the influence of DNA elastic properties on the natural frequency of microbeam. The governing equation of the DNA-microbeam system will be established by using the energy method, and the first-order variation of the relevant generalized Hamiltonian function is written as

$$\delta \int_{t_1}^{t_2} \int_0^l (T - \Pi) dx dt + \int_{t_1}^{t_2} \int_0^l \delta V dx dt = 0 \tag{4}$$

where T, $\prod$, and V, respectively, represent the kinetic energy per unit axial length, total elastic potential energy of the DNA-microbeam system, and external work per unit axial length; t_1 and t_2 are different moments.

As for the dynamic response of a hinged–hinged beam, the kinetic energy per unit axial length, T, can be written as

$$T = (m + \Delta m)(\partial w/\partial t)^2/2, \tag{5}$$

where m and Δm represent the linear mass density of the substrate and the DNA film, respectively.

Considering the surface stress σ_p as a symmetric external load along the surface of the substrate, the external work per unit axial length can be written as

$$V = \sigma_p h_p b (\partial w/\partial x)^2. \tag{6}$$

The total elastic potential energy of the DNA-microbeam system, $\prod$, includes three parts: The elastic potential energy stored in the substrate, W_s, the effective elastic potential energy of the top DNA film, $W_{p,top}$, and that of the bottom DNA film, $W_{p,bot}$, i.e.,

$$\Pi = W_s + W_{p,bot} + W_{p,top}, \tag{7}$$

in which

$$\begin{aligned} W_s &= bl\int_{-h_s/2}^{h_s/2} E_s\varepsilon^2 dz/2 \\ W_{p,\text{bot}} &= bl\int_{h_s/2}^{h_s/2+h_p} (\sigma_p\varepsilon + E_s\varepsilon^2/2)dz \\ W_{p,\text{top}} &= bl\int_{-h_s/2-h_p}^{-h_s/2} (\sigma_p\varepsilon + E_s\varepsilon^2/2)dz \end{aligned}$$

where the bending strain ε can be described by Zhang's two variable method [27], i.e., $\varepsilon = \varepsilon_0 - \kappa z$, where κ is the curvature of the neutral axis and ε_0 is the normal strain along the x-direction at $z = 0$. The effective elastic potential energies of the adsorbed DNA films are estimated by using Equations (1)–(3).

Substituting Equations (5)–(7) into Equation (4), the vibrational differential equation is obtained as

$$(E_sI_s + \Delta EI)\partial^4 w/\partial x^4 + 2\sigma_p h_p b\partial^2 w/\partial x^2 = (m + \Delta m)\partial^2 w/\partial t^2 \tag{8}$$

where $m = \rho b h_s$, ρ, and E_sI_s are, respectively, the effective linear mass density, the mass density, and the stiffness of the substrate; b is the beam width; $\Delta m \approx 2\eta b N \times 1.6 \times 10^{-21}/1600$ kg is the effective mass of the DNA film per unit axial length of the beam [28]; $\Delta EI = E_p b I_{u,\text{bot2}} + E_p b I_{u,\text{top2}}$ is the additional stiffness induced by DNA adsorptions, and $I_{u,\text{bot2}} = I_{u,\text{top2}} = \int_{h_s/2}^{h_s/2+h_p} z^2 dz$. Note that the effective stiffness could reduce to that of Eom et al. [13], Wang and Feng [29], and Lu et al. [30] in the case of tiny film thickness.

The separation variable method is used to solve Equation (8). Assume $w(x, t) = \Phi(x)q(t)$, where $\Phi(x)$ is the modal function and $q(t)$ is the time domain function. To illustrate the surface effects, here only the hinged–hinged microbeam is considered. Substituting the above solution form into Equation (8) yields the following i-th mode natural frequency of the beam after DNA adsorptions, i.e.,

$$\begin{aligned} p_i &= p_i^0\sqrt{\alpha_1\alpha_2\alpha_3} \\ \alpha_1 &= 1 + \Delta EI/E_sI_s, \\ \alpha_2 &= 1 + 2\sigma_p h_p b l^2/[i^2\pi^2(E_sI_s + \Delta EI)], \\ \alpha_3 &= 1 - \Delta m_{\text{DNA}}/(\rho a h_s + \Delta m_{\text{DNA}}), \end{aligned} \tag{9}$$

where p_i^0 is the i-th mode natural frequency without surface effect; α_1, α_2, and α_3 are the dimensionless parameters standing for the effects of surface stiffness, stress–stiffness coupling, and additional mass, respectively. Obviously, the above three effects are closely related to the geometric and elastic properties of adsorbed DNA films and the substrate.

To summarize, different microscopic interactions of surface molecules may endow DNA films with totally different mechanical properties, which are closely relevant to the complex detection signals of DNA-microbeams. With the above analytical model, we can quantify these multiscale correlations between macroscopic detection signals and surface elastic properties of the adsorbed film induced by microscopic molecular interactions.

3. Results and Discussion

In computation, dsDNA nucleotide number is taken as $N = 25$, the substrate size $l = 9$ μm, and $b = 0.4$ μm for dynamic analyses of a hinged–hinged microbeam. Due to the length-to-width ratio of the substrate, the biaxial modulus is taken as $E_s/(1-\mu_s)$, where elastic modulus $E_s = 180$ GPa, and Poisson's ratio $\mu_s = 0.27$. The parameters in Equation (3) for dsDNA in 0.1 M NaCl solution is taken as: $a_0 = 0.41 \times 10^{-9}$ J/m, $b_0 = 1.1 \times 10^{-7}$ J/m, $c_0 = 0.8$, $\lambda_D = 0.974$ nm, and $\lambda_H = 0.288$ nm [25]. Substituting the experimental data on ΔG_{rep} and $\Delta G_{\text{att}} = W_{b2} - \Delta G_{\text{rep}}$ of Todd et al. [4] into Equation (2), we could obtain the prefactors C_R and C_A of ΔG_{rep} and ΔG_{att} for dsDNA, and the related parameters in different salt solutions are shown in Table 1. Here, 12.3 pN = $1k_BT/a$, in which the nucleotide length $a = 0.34$ nm. According to the previous osmotic pressure experiments [31], the interaxial spacing of dsDNA inside virus is about 2.6 nm, so the packing density can approximately reach 1.7×10^{17} chains/m^2 for the hexagonal packing pattern.

Table 1. Experimental data of double-stranded DNA (dsDNA) at different salt solutions [4] and the solved prefactors.

	d, Å	aW_{b2}, k_BT/a	$a\Delta G_{rep}$, k_BT/a	C_A, MPa	C_R, MPa
$Co(NH_3)_6Cl_3$	27.75	−0.21	0.17	755.83	303, 444
Spermine	28.15	−0.33	0.29	945.89	508, 714
sp^{6+}	27.65	−0.38	0.39	1503.26	668, 743

First, we will study the variation of surface elastic properties of adsorbed dsDNA films and its mechanism induced by micro-interactions. By using Equation (1), the variation tendencies of surface stress with the packing density in several salt solutions are compared in Figure 2a. By convention, the positive value represents the compressive stress while the negative value represents the tensile one. In NaCl solution, the surface stress always behaved compressive and its value increased with the enhancement of the packing density. In addition, Figure 2b shows that the collected contributions of electrostatic energy, hydration energy, and configurational entropy led to the variation of surface stress, which was also the deformation mechanism of micro-beam sensor in NaCl solution.

Whereas in multivalent solutions (sp^{6+} and spermine), the surface stress exhibited different trends with nonmonotonic variations, as shown in Figure 2a. Taking sp^{6+} as an example, when the packing density $\eta < 1.45 \times 10^{17}$ chain/m^2, the surface stress behaved tensile and this revealed the possible physical mechanism of the condensation found in organism induced by the attractive interactions between DNA helices; when $\eta \approx 1.2 \times 10^{17}$ chain/m^2, the tensile surface stress reached its maximum value, which provided us an opportunity to prepare a more sensitive sensor by the directed-assembled technique; when $\eta \approx 1.45 \times 10^{17}$ chain/m^2, the surface stress turned to be zero and the sensor might have lost any signals, and this is the most miserable situation in biodetections; when $\eta > 1.45 \times 10^{17}$ chain/m^2, the surface stress behaved compressive inversely and this indicates the dominance of the repulsive interactions between DNA helices. Physically speaking, the competition between repulsive and attractive part of free energy make the surface stress changing from tensile to compressive, and this also interprets the mechanism of microbeam sensor deformation in sp^{6+} solution. As shown in Figure 2c, at a relatively low packing density, the dominance of contribution of the attractive part of the free energy resulted in tensile surface stresses; with the increase in the packing density, the repulsive part of the free energy gradually became more critical and eventually resulted in compressive surface stresses. However, the discrepancy in $Co(NH_3)_6Cl_3$ came into sight. While in $Co(NH_3)_6Cl_3$ solution, the surface stress will always be tensile. In addition, given the parameters exactly the same as in the experiment, the magnitude of the tensile surface stress was about 1 MPa, and it has the same order with Todd's experimental result of osmotic pressure among DNA molecules, i.e., $\Pi \in (0.1$ MPa, 10 MPa$)$ [4].

Also, by using Equation (1), the variation tendencies of elastic modulus with the packing density have been studied. Figure 3a shows the elastic moduli of dsDNA films in various salt solutions. As we can see, with the similar tendency of the surface stress, the elastic modulus of the DNA film in NaCl solution always behaved positive and it increased with the enhancement of the packing density. As shown in Figure 3b, in NaCl solution, the collected contributions of electrostatic energy, hydration energy, and configurational entropy to the surface stress lead to the variation of elastic modulus at different packing densities. Nevertheless, the elastic modulus in a multivalent solution (sp^{6+}, spermine, and $Co(NH_3)_6Cl_3$) was negative at a relatively low DNA packing density, whereas it turned positive at a relatively high density. In addition, the elastic moduli in multivalent solutions were about one order of magnitude lower than that in NaCl solution. However, the critical packing densities in Figure 2a and 3a are different. For example, in sp^{6+} solutions, the DNA elastic modulus was negative when the packing density $\eta < 1.14 \times 10^{17}$ chain/m^2, and became almost zero when η reached

1.14×10^{17} chain/m^2, then turned positive when $\eta > 1.14 \times 10^{17}$ chain/m^2. Also, there was a critical packing density for the negative modulus at $\eta \approx 0.95 \times 10^{17}$ chain/m^2. Furthermore, Figure 3c shows that the competition between the repulsive part and attractive part of free energy lead to the non-monotonic variation of elastic modulus. In addition, the magnitude of the elastic modulus of the DNA film in 0.1 M NaCl solution was about 0.1~100 MPa, which is similar to Zhang's theoretical prediction [22], and slightly smaller than Legay's (50 mM NaCl solution) [32], due to different salt concentrations and packing conditions as well as the inherent deficiency of AFM-based nano-indentation detection. What is more, our simulation showed a consistent monotonic trend with that of Domínguez's theoretical predictions and approaches the order of their AFM experiment results [17]. It should be mentioned that negative elastic modulus is unstable in nature, however can be stabilized by lateral constraint [33,34]. As for the DNA film in the microbeam-based biosensor, it was actually pre-stretched during the immobilization process, namely restrained by the substrate. Figuratively speaking, imagining the DNA film as a pre-stretched spring, it is surely unstable without lateral constraint. When we dismiss the constraint and apply a tiny lateral tensile stress far less than the residual stress induced by pre-stretching, which is insufficient to remain the stable state, the pre-stretched spring will obviously be compressed and consequently behaves a negative modulus.

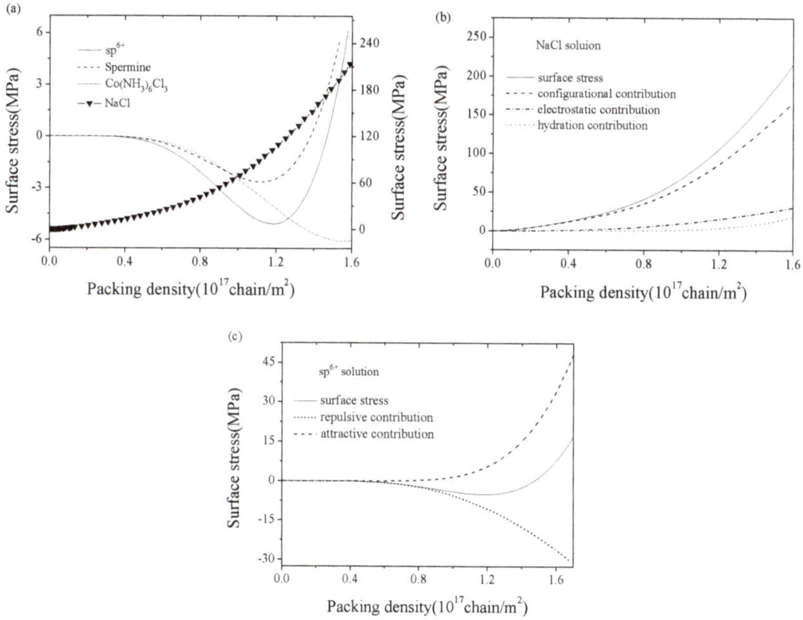

Figure 2. (**a**) Surface stress variation of dsDNA films with packing density in sp^{6+}, spermine, $Co(NH_3)_6Cl_3$, and NaCl solutions. The left longitudinal axis is related to DNA in multivalent solutions and the right one is related to DNA in NaCl solution. By convention, the positive value represents the compressive stress while the negative value means the tensile one. (**b**) Contributions of electrostatic energy, hydration energy, and configurational entropy to the surface stress in NaCl solution. (**c**) Contributions of the repulsive and attractive parts of free energy to the surface stress in sp^{6+} solution.

Next, by using Equation (9), we will study the variation of the natural frequency shift of a hinged–hinged microbeam induced by dsDNA adsorptions and its mechanism related to surface properties. As we can see from Equation (9), the natural frequency shift was the result of the competition between effects of surface stiffness, stress–stiffness coupling, and additional mass (α_1, α_2, α_3), which is closely related to the elastic properties of adsorbed films induced by micro-interactions as well as the elastic and geometric properties of the substrate. It can be learned from the above discussions that, given the packing density η = 1.2 × 10^{17} chain/m^2, the surface stress of dsDNA film will always behave compressive in NaCl solution or tensile in sp^{6+} solution, respectively. Considering the boundary constraints, obviously the substrate will be compressed in NaCl solution and stretched in sp^{6+} solution, respectively. Once the elastic moduli and surface stress of the adsorbed dsDNA film are known, the dynamic detection signals of dsDNA-microbeam could be easily obtained.

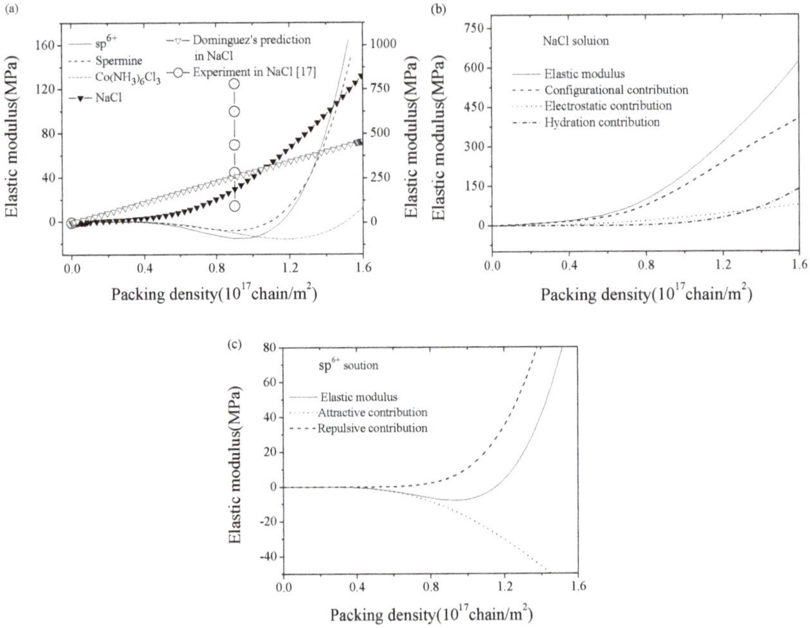

Figure 3. (**a**) Theoretical elastic modulus variation (lines) of dsDNA films with packing density in sp^{6+}, spermine, $Co(NH_3)_6Cl_3$, NaCl solutions, and Domínguez's AFM experiment results [17] (circles) in NaCl solution. The left longitudinal axis is related to DNA in multivalent solutions and the right one is related to DNA in NaCl solution. (**b**) Contributions of electrostatic energy, hydration energy, and configurational entropy to the elastic modulus in NaCl solution. (**c**) Contributions of the repulsive and attractive part of free energy to the elastic modulus in sp^{6+} solution.

Figure 4 shows the first-order natural frequency shift of the hinged–hinged microbeam with the variation in the absolute value of film-to-substrate thickness ratio (i.e., $r = |h_p/h_s|$) and modulus ratio (i.e., $g = |E_p/E_s|$). First, as shown in Figure 4, the first-order natural frequency shift due to dsDNA adsorptions was mostly negative in NaCl solution and positive in sp^{6+} solution. Similar behavior has been discovered in Karabalin's surface stress loaded beam experiments (beam length: 6 to 10 μm; width: 0.6 to 1 μm; thickness: 0.015 to 0.028 μm; Poisson's ratio: 0 to 0.49) [35]) and Lachut's analytical predictions [36]. Second, the amplitudes of the natural frequency shift in both solutions showed the similar tendency, namely, enhancing with the increase of the absolute value of film-to-substrate thickness ratio or modulus ratio. Actually, as shown in Figure 4, when the parameter values were relatively large, the stress–stiffness coupling effect α_2 dominated the value of natural frequency shift. Taking sp^{6+} solution as example, given $r = g = 0.04$, the contributions of α_1, α_2, α_3 to the first-order natural frequency shift were, respectively, −0.52%, 44.5%, and −0.6%, so the positive effect of the stress–stiffness coupling determined the upward trend of natural frequency shift.

Third, an anomalous invalid region is observed in Figure 4. Note that the DNA film in NaCl solution is in a repulsion-dominated state, and the microbeam vibrates in different modes depending on the specific experiment conditions:

(i) When the relation between the modulus ratio and the thickness ratio satisfies the following relation, $g \leq 9.016 \times 10^{-6}/(6.22r^3 - 1.08 \times 10^{-4}r^2 - 5.41 \times 10^{-5}r)$, the microbeam vibrates in a linear phase, in which the frequency shift of a periodic vibration could be taken as an indication of DNA adsorptions;

(ii) When their relation satisfies the following relation, $g > 9.016 \times 10^{-6}/(6.22r^3 - 1.08 \times 10^{-4}r^2 - 5.41 \times 10^{-5}r)$, the microbeam vibrates in a non-periodic way, which means a dynamic instability region (i.e., the anomalous blank area in Figure 4 appears.

It can be seen from the linear analytical solution to Equation (8) that, when the parameters locate at the condition (ii), the additional mass-relevant coefficient α_3 is always greater than zero, the competition between the surface stress effect and the stiffness effect makes the signs of α_1 and α_2 opposite, and this means $p_i^2 < 0$, so its corresponding temporal-domain equation, $\ddot{q}(t) + p_i^2 q(t) = 0$, has a nonperiodic solution with $q_1(t) = -c_1 p_i^{-2} e^{-p_i^2 t} + c_2$; here, c_1 and c_2 are determined by the initial conditions. In other words, the motion increases exponentially. This is totally different from the linear periodic motion with $q_2(t) = c\sin(p_i t + \theta)$ when $p_i^2 > 0$, where c is also determined by the initial conditions. The restriction on the linear periodic motion endows the linear vibration natural frequency shift only with an upper limit of 100% in NaCl solution. The instability indicates the occurrence of a sensitive interval in which DNA adsorptions induce a drastic natural frequency shift even with a tiny variation of the detection conditions. Whereas the appearance of dynamic instability at condition (ii) will cause a sudden jump of vibration mode from stability to instability at the critical condition, and this means a relatively large deformation for the beam. In these cases, this dynamic instability might provide us a potential method to develop a ultra-highly sensitive detection method through the linear vibration-based material parameter controlling or a new nonlinear vibration-based technology in the future [37].

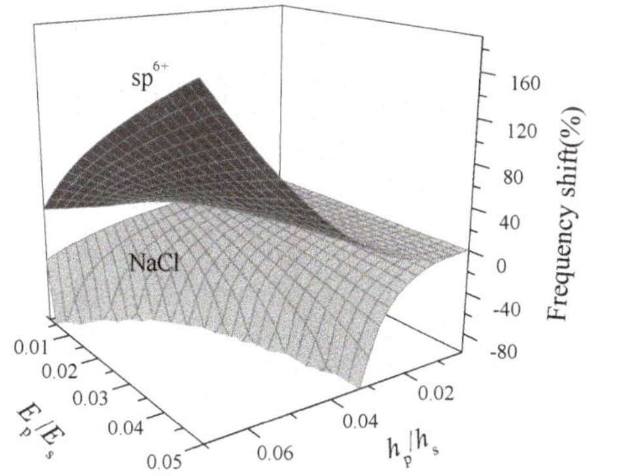

Figure 4. The first-order natural frequency shift of a hinged–hinged dsDNA-microbeam with the variation in the absolute value of film-to-substrate thickness ratio (i.e., $r = |h_p/h_s|$) and modulus ratio (i.e., $g = |E_p/E_s|$) in sp^{6+} and NaCl solutions when the packing density $\eta = 1.2 \times 10^{17}$ chain/m^2.

However, unlike the seemingly monotonicity observed in the global view as shown in Figure 4, when the parameter value was taken relatively small, the non-monotonic behavior came into sight in the local zoom view, as shown in Figure 5. Taking the detection of attraction-dominated films in sp^{6+} solution as an example, given the modulus ratio $g = 0.04$ in Figure 5a, when the thickness ratio $r < 0.00572$, the frequency shift was negative, and became almost zero when r reached 0.00572, then turned positive when $r > 0.00572$. Also, there was a critical value for the negative frequency shift at $r \approx 0.00347$. However, given the modulus ratio $g = 0.1$, the frequency shift tendency in the detection of repulsion-dominated films as shown in Figure 5b was totally different from that of attraction-dominated films, and with the increase in thickness ratio the shift turned from positive to negative. In addition, the non-monotonic behavior observed in Figure 5b was negligible when the modulus ratio was relatively small (e.g., $g = 0.04$). The variation between positive and negative frequency shift has been found in DNA hybridization experiments by Zheng et al. [38], and similar anomalous non-monotonic tendencies have been found in the study of alkanethiol adsorption by Tamayo et al. (beam material: Si; critical thickness ratio h_p/h_s approximates to 0.15) [15] and Au adsorption by Lee et al. (beam material: lead zirconate titanate (PZT); critical thickness ratio h_p/h_s approximates to 0.000445) [14] when the adsorption layer is relatively thin compared with the substrate.

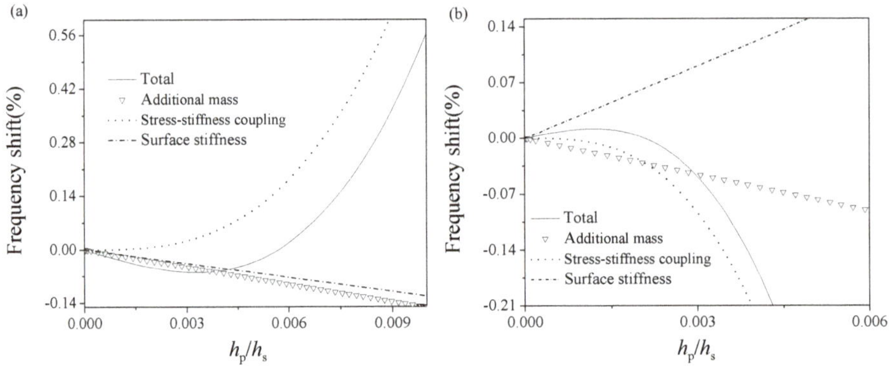

Figure 5. The first-order natural frequency shift of a hinged–hinged dsDNA-microbeam and contributions of surface stiffness (α_1), stress–stiffness coupling (α_2), and additional mass (α_3) effects with the variation in the absolute value of film-to-substrate thickness ratio (i.e., $r = |h_p/h_s|$) in sp^{6+} and NaCl solutions when the packing density $\eta = 1.2 \times 10^{17}$ chain/m^2. (**a**) sp^{6+} solution, film-to-substrate modulus ratio $g = |E_p/E_s| = 0.04$; (**b**) NaCl solution, $g = |E_p/E_s| = 0.1$.

The physical mechanism underlying these non-monotonic responses of the microbeam to different detection conditions can be interpreted by the present multiscale analytical model. As we can see from Figure 5a, during the detection of the attraction-dominated films in sp^{6+} solution, the effects of additional mass α_3 and surface stiffness α_1 always behaved negatively whereas the effect of stress–stiffness coupling α_2 behaved positively, which means that the stress–stiffness coupling effect dominated when the thickness ratio $r > 0.00572$, whereas both additional mass and surface stiffness effect played an essential role when $r < 0.00572$. In a word, the frequency shift of the microbeam is the result of the competition of the above-mentioned three effects closely related to the elastic and geometric properties of the adsorbed film and the substrate. Different surface elastic properties of repulsion-dominated films in NaCl solution leads to a totally different tendency in the frequency shift. These conclusions further verified the previous conclusion based on experimental observations that the stress–stiffness coupling effect becomes more dominant with the increase in the absolute value of film-to-substrate thickness ratio [14]. In addition, these non-monotonic variations and totally different responses in frequency shifts during the detections of attraction- or repulsion-dominated dsDNA films provide us an alternative perspective to promote the sensitivity of surface-effect-based biosensors.

It should be mentioned that, in the case of detecting the adsorbed DNA film with an anomalous negative elastic modulus, if we mistake it as a general material with a positive modulus, this might induce a large prediction error. Taking the dynamic signals of DNA films with $h_p/h_s = 0.003$ in sp^{6+} solution as example, the elastic modulus of DNA film is about -1 MPa. As shown in Figure 5a, the original prediction of the frequency shift considering the negative elastic modulus was about -0.059%. However, if we take the elastic modulus as 1 MPa inversely, the frequency shift will be mistakenly estimated as 0.01%, and the relative prediction error between these two predictions is about 100%. In a word, this unneglectable prediction error indicates that the anomalous negative elastic modulus of the adsorbed DNA film has great influence on microbeam-based biodetection signals.

4. Conclusions

In this paper, we characterized the elastic properties of dsDNA films and established a multiscale analytical model to describe the relationship between the surface mechanical properties of DNA self-assembled 2D films and the detection signals of DNA-microbeam systems. The related predictions agree well with the AFM indentation experiment [17] and microbeam vibration experiment [14,35]. Analytical results show that the microscale attractive interactions between DNA chains will lead to anomalous negative elastic moduli and tensile surface stresses, and the occurrence of this tensile surface stress for the attraction-dominated DNA film reveals the possible physical mechanism of the condensation found in organism. In addition, the dynamic analysis of a hinged–hinged microbeam in multivalent salt solutions suggests that, despite the non-monotonic tendency of frequency shift when the absolute value of film-to-substrate thickness ratio is relatively small, above a critical film-to-substrate thickness ratio, an attraction-dominated film could always induce a positive natural frequency shift, totally different from the detection signal for a repulsion-dominated DNA film. These insights emphasize the importance of the stress–stiffness coupling effect in dynamic responses and provide us an alternative perspective to promote the sensitivity of surface-effect-based biosensor. What is more important, during the detection of a repulsion-dominated DNA film, dynamic instability appears after the critical conditions, which brings about a jump of vibration mode from stable to instable states with a relatively large displacement of a microbeam, and this indicates the existence of a sensitive interval in which DNA adsorptions will induce a drastic natural frequency shift even with a tiny variation of the detection conditions. In these cases, this dynamic instability might provide us a potential method to develop an ultra-highly sensitive detection method through the linear vibration-based material parameter controlling or a new nonlinear vibration-based technology in the future.

Author Contributions: N.Z. designed the research; J.W. contributed the calculation and data analysis; J.W. and Y.Z. wrote the paper under the supervision of N.Z.

Funding: This research was funded by the National Natural Science Foundation of China [Grant Nos. 11772182, 11272193, 10872121].

Acknowledgments: The authors gratefully acknowledge the financial support from the National Natural Science Foundation of China [Grant Nos. 11772182, 11272193, 10872121].

Conflicts of Interest: The authors declare no conflict of interest.

References

1. Bloomfield, V.A. DNA condensation by multivalent cations. *Biopolymers* **1997**, *44*, 269–282. [CrossRef]
2. Cortini, R.; Caré, B.R.; Victor, J.; Barbi, M. Theory and simulations of toroidal and rod-like structures in single-molecule DNA condensation. *J. Chem. Phys.* **2015**, *142*, 105102. [CrossRef]
3. Li, G.Y.; Guan, R.L.; Ji, L.N.; Chao, H. DNA condensation induced by metal complexes. *Coord. Chem. Rev.* **2014**, *281*, 100–113. [CrossRef]
4. Todd, B.A.; Parsegian, V.A.; Shirahata, A.; Thomas, T.J.; Rau, D.C. Attractive forces between cation condensed DNA double helices. *Biophys. J.* **2008**, *94*, 4775–4782. [CrossRef] [PubMed]
5. Mertens, J.; Tamayo, J.; Kosaka, P.; Calleja, M. Observation of spermidine-induced attractive forces in self-assembled monolayers of single stranded DNA using a microcantilever sensor. *Appl. Phys. Lett.* **2011**, *98*, 153704. [CrossRef]
6. Montasser, I.; Coleman, A.W.; Tauran, Y.; Perret, G.; Jalabert, L.; Collard, D.; Kim, B.J.; Tarhan, M.C. Direct measurement of the mechanism by which magnesium specifically modifies the mechanical properties of DNA. *Biomicrofluidics* **2017**, *11*, 051102. [CrossRef]
7. Jeltsch, A.; Maschke, H.; Selent, U.; Wenz, C.; Köhler, E.; Connolly, B.A.; Thorogood, H.; Pingoud, A. DNA binding specificity of the EcoRV restriction endonuclease ss increased by Mg2+ binding to a metal ion

binding site distinct from the catalytic center of the Enzyme. *Biochemistry* **1995**, *34*, 6239–6246. [CrossRef] [PubMed]
8. Chen, C.S.; Chou, C.C.; Chang, S.W. Multiscale analysis of adsorption-induced surface stress of alkanethiol on microcantilever. *J. Phys. D Appl. Phys.* **2013**, *46*, 035301. [CrossRef]
9. Mathew, R.; Sankar, A.R. Design of a triangular platform piezoresistive affinity microcantilever sensor for biochemical sensing applications. *J. Phys. D Appl. Phys.* **2015**, *48*, 205402. [CrossRef]
10. Zhang, G.M.; Zhao, L.B.; Jiang, Z.D.; Yang, S.M.; Zhao, Y.L.; Huang, E.; Hebibul, R.; Wang, X.P.; Liu, Z.G. Surface stress-induced deflection of a microcantilever with various widths and overall microcantilever sensitivity enhancement via geometry modification. *J. Phys. D Appl. Phys.* **2011**, *44*, 425402. [CrossRef]
11. Stachowiak, J.C.; Yue, M.; Castelino, K.; Chakraborty, A.; Majumdar, A. Chemomechanics of surface stresses induced by DNA hybridization. *Langmuir* **2006**, *22*, 263–268. [CrossRef]
12. Biswal, S.L.; Raorane, D.; Chaiken, A.; Majumdar, H.B.A. Nanomechanical Detection of DNA Melting on Microcantilever Surfaces. *Anal. Chem.* **2006**, *78*, 7104–7109. [CrossRef]
13. Eom, K.; Kwon, T.Y.; Yoon, D.S.; Lee, H.L.; Kim, T.S. Dynamical response of nanomechanical resonators to biomolecular interactions. *Phys. Rev. B* **2007**, *76*, 113408. [CrossRef]
14. Lee, J.H.; Hwang, K.S.; Yoon, D.S.; Kim, H.; Song, S.H.; Kang, J.Y.; Kim, T.S. Anomalous resonant frequency changes in piezoelectric microcantilevers by monolayer formation of Au films. *Appl. Phys. Lett.* **2011**, *99*, 143701. [CrossRef]
15. Tamayo, J.; Ramos, D.; Mertens, J.; Calleja, M. Effect of the adsorbate stiffness on the resonance response of microcantilever sensors. *Appl. Phys. Lett.* **2006**, *89*, 224104. [CrossRef]
16. Wu, J.Z.; Zhou, M.H.; Zhang, N.H. The effect of microscopic attractive interactions on piezoelectric coefficients of nanoscale DNA films and its resultant mirocantilever-based biosensor signals. *J. Phys. D Appl. Phys.* **2017**, *50*, 415403. [CrossRef]
17. Domínguez, C.M.; Ramos, D.; Mendieta-Moreno, J.I.; Fierro, J.L.G.; Mendieta, J.; Tamayo, J.; Calleja, M. Effect of water-DNA interactions on elastic properties of DNA self-assembled monolayers. *Sci. Rep.* **2017**, *7*, 536. [CrossRef] [PubMed]
18. Shu, W.M.; Laue, E.D.; Seshia, A.A. Investigation of biotin–streptavidin binding interactions using microcantilever sensors. *Biosens. Bioelectron.* **2007**, *22*, 2003–2009. [CrossRef]
19. Shu, W.M.; Laurenson, S.; Knowles, T.P.J.; Ferrigno, P.K.; Seshia, A.A. Highly specific label-free protein detection from lysed cells using internally referenced microcantilever sensors. *Biosens. Bioelectron.* **2008**, *24*, 233–237. [CrossRef]
20. Zhou, M.H.; Meng, M.L.; Zhang, C.Y.; Li, X.B.; Wu, J.Z.; Zhang, N.H. The pH-dependent elastic properties of nanoscale DNA films and the resultant bending signals for microcantilever biosensors. *Soft Matter* **2018**, *14*, 3028–3039. [CrossRef] [PubMed]
21. Zhang, N.H.; Shan, J.Y. An energy model for nanomechanical deflection of cantilever-DNA chip. *J. Mech. Phys. Solids* **2008**, *56*, 2328–2337. [CrossRef]
22. Zhang, N.H.; Meng, W.L.; Tan, Z.Q. A multi-scale model for the analysis of the inhomogeneity of elastic properties of DNA biofilm on microcantilevers. *Biomaterials* **2013**, *34*, 1833–1842. [CrossRef]
23. Hagan, M.F.; Majumdar, A.; Chakraborty, A.K. Nanomechanical Forces Generated by Surface Grafted DNA. *J. Phys. Chem. B* **2002**, *106*, 10163–10173. [CrossRef]
24. Rekesh, D.; Lyubchenko, Y.; Shlyakhtenko, L.S.; Lindsay, S.M. Scanning tunneling microscopy of mercapto-hexyl-oligonucleotides attached to gold. *Biophys. J.* **1996**, *71*, 1079–1086. [CrossRef]
25. Strey, H.H.; Parsegian, V.A.; Podgornik, R. Equation of State for DNA Liquid Crystals: Fluctuation Enhanced Electrostatic Double Layer Repulsion. *Phys. Rev. Lett.* **1997**, *78*, 895–898. [CrossRef]
26. Ambia-Garrido, J.; Vainrub, A.; Pettitt, B.M. A model for structure and thermodynamics of ssDNA and dsDNA near a surface: A coarse grained approach. *Comput. Phys. Commun.* **2010**, *181*, 2001–2007. [CrossRef]
27. Zhang, N.H.; Xing, J.J. An alternative model for elastic bending deformation of multilayered beams. *J. Appl. Phys.* **2006**, *100*, 103519. [CrossRef]
28. Ilic, B.; Yang, Y.; Aubin, K.L.; Reichenbach, R.; Krylov, S.; Craiqhead, H.G. Enumeration of DNA molecules bound to a nanomechanical oscillator. *Nano Lett.* **2005**, *5*, 925–929. [CrossRef]
29. Wang, G.F.; Feng, X.Q. Effects of surface elasticity and residual surface tension on the natural frequency of microbeams. *Appl. Phys. Lett.* **2007**, *90*, 231904. [CrossRef]

30. Lu, P.; Lee, H.P.; Lu, C.; O'Shea, S.J.O. Surface stress effects on the resonance properties of cantilever sensors. *Phys. Rev. B* **2005**, *72*, 085405. [CrossRef]
31. Yasar, S.; Podgornik, R.; Valle-Orero, J.; Johnson, M.R.; Parsegian, V.A. Continuity of states between the cholesteric → line hexatic transition and the condensation transition in DNA solutions. *Sci. Rep.* **2014**, *4*, 6877. [CrossRef]
32. Legay, G.; Finot, E.; Meunier-Prest, R.; Cherkaoui-Malki, M.; Latruffe, N.; Dereux, A. DNA nanofilm thickness measurement on microarray in air and in liquid using an atomic force microscope. *Biosens. Bioelectron.* **2005**, *21*, 627–636. [CrossRef] [PubMed]
33. Lakes, R.S.; Rosakis, P.; Ruina, A. Microbuckling instability in elastomeric cellular solids. *J. Mater. Sci.* **1993**, *28*, 4667–4672. [CrossRef]
34. Lakes, R.S. Extreme damping in composite materials with a negative stiffness phase. *Phys. Rev. Lett.* **2001**, *86*, 2897–2900. [CrossRef] [PubMed]
35. Karabalin, R.B.; Villanueva, L.G.; Matheny, M.H.; Sader, J.E.; Roukes, M.L. Stress-induced variations in the stiffness of micro- and nanocantilever beams. *Phys. Rev. Lett.* **2012**, *108*, 236101. [CrossRef]
36. Lachut, M.J.; Sader, J.E. Effect of surface stress on the stiffness of thin elastic plates and beams. *Phys. Rev. B* **2012**, *85*, 085440. [CrossRef]
37. Kozinsky, I.; Postma, H.W.C.; Kogan, O.; Husain, A.; Roukes, M.L. Basins of attraction of a nonlinear nanomechanical resonator. *Phys. Rev. Lett.* **2007**, *99*, 207201. [CrossRef] [PubMed]
38. Zheng, S.; Choi, J.H.; Lee, S.M.; Hwang, K.S.; Kim, S.K.; Kim, T.S. Analysis of DNA hybridization regarding the conformation of molecular layer with piezoelectric microcantilevers. *Lab Chip* **2011**, *11*, 63–69. [CrossRef]

nanomaterials

MDPI

Article

Self-Assembled Triphenylphosphonium-Conjugated Dicyanostilbene Nanoparticles and Their Fluorescence Probes for Reactive Oxygen Species

Wonjin Choi [1,†], Na Young Lim [1,†], Heekyoung Choi [1], Moo Lyong Seo [1,*], Junho Ahn [2,*] and Jong Hwa Jung [1,*]

1 Department of Chemistry and Research Institute of Natural Science, Gyeongsang National University, Jinju 52828, Korea; cwj1685@gnu.ac.kr (W.C.); skdud325@gnu.ac.kr (N.Y.L.); smile377@gnu.ac.kr (H.C.)
2 Composites Research Division, Korea Institute of Materials Science, Changwon 51508, Korea
* Correspondence: mlseo@gnu.ac.kr (M.L.S.); junho2587@kims.re.kr (J.A.); jonghwa@gnu.ac.kr (J.H.J.); Tel.: +82-55-772-1488 (J.H.J.)
† These authors contributed equally to this work.

Received: 23 November 2018; Accepted: 10 December 2018; Published: 12 December 2018

Abstract: We report self-assembled novel triphenylphosphonium-conjugated dicyanostilbene- based as selective fluorescence turn-on probes for $^1O_2$ and ClO^-. Mono- or di-triphenylphosphonium-conjugated dicyanostilbene derivatives **1** and **2** formed spherical structures with diameters of ca. 27 and 56.5 nm, respectively, through π-π interaction between dicyanostilbene groups. Self-assembled **1** showed strong fluorescent emission upon the addition of $^1O_2$ and ClO^- compared to other ROS (O_2^-, $^\bullet OH$, NO, TBHP, H_2O_2, GSH), metal ions (K^+, Na^+), and amino acids (cysteine and histidine). Upon addition of $^1O_2$ and ClO^-, the spherical structure of **1** changed to a fiber structure (8-nm wide; 300-nm long). Upon addition of $^1O_2$ and ClO^-, the chemical structural conversion of **1** was determined by FAB-Mass, NMR, IR and Zeta potential analysis, and the strong emission of the self-assembled **1** was due to an aggregation-induced emission enhancement. This self-assembled material was the first for selective ROS as a fluorescence turn-on probe. Thus, a nanostructure change-derived turn-on sensing strategy for $^1O_2$ or ClO^- may offer a new approach to developing methods for specific guest molecules in biological and environmental subjects.

Keywords: dicyanostilbene; triphenylphosphonium; self-assembly; ROS detection

1. Introduction

Generally, self-assembly of aggregation-induced emission (AIE)-active molecules in aqueous solution depends on intermolecular hydrogen bonding, π-π stacking, and hydrophobic interactions that form nanoscale architectures, such as spheres, rods, and fibers, with fluorescence enhancement [1,2]. Nanostructures formed via molecular self-assembly involving stimuli-responsive properties have great potential for biological and environmental applications [3–6]. Despite the utilization of dual effects on fluorescence turn-on originating from intramolecular electron transfer and intermolecular self-assembly [7], fluorogenic or chromogenic sensing of biologic or environmental species with selective turn-on detection have rarely been reported due to the difficulty associated with the design and synthesis of fluorescence probes. Strong interactions between acidic protons and solvent molecules cause act as a fluorescence quenching factor by photoelectron transfer (PET). Therefore, the self-assembled AIE fluorescence probes with hydrophobic positive charge are useful as sensing materials in biological field [8–10].

Studies related to optical detection using chromogenic or fluorogenic chemoprobes have been performed for biological and environmental applications of singlet oxygen ($^1O_2$) and hypochlorite

(ClO^-), such as the main component of a cleaning agent for industrial wastewater and cancer treatments that destroy tumor cells [11–13]. Fluorescent probes that demonstrate rapid response, high sensitivity, and technical simplicity are attractive tools for analyte monitoring [14], and fluorescent probes with a reaction-induced signal have been designed to detect a specific signal for singlet oxygen or hypochlorite [15]. In addition, a variety of fluorescent probes for ROS detection have been reported [11,16]. For example, a probe for singlet oxygen primarily produced detection signals by the addition reaction of singlet oxygen to anthracene or the rhodamine backbone [17,18]. In addition, an Ir (III) complex-linked coumarin 314 derivative has been used to identify a ratiometric signal through $^1O_2$-mediated abstraction of the α-H from the tertiary amine [19]. In a probe for hypochlorite detection, fluorescent probe reactions with hypochlorite, such as oxidative cleavage reaction of double bonds (C=C, C=S, C=N, and N=N) or oxidative–hydrolysis reaction of amide, diphenyl ether, and thioether, produced detection signals from the reaction product [20].

In particular, depending on the self-assembly conditions, amphiphilic molecules with AIE have attracted significant attention due to the ease of controlling the fluorescence of an aggregate, which facilitates the construction of fluorescent turn off-on systems [21]. To obtain fluorescence turn-on signals by the reaction of a probe with an analyte, AIE molecules, such as tetraphenylethylene (TPE), are typically applied for ROS detection [22]. Dicyanodistyrylbenzenes are p-conjugated molecules with various optical properties, such as AIE in emission properties and tunable luminescence emission [23]. Compared to homologous a-cyanostilbenes, emission spectra occur at higher wavelengths due to their longer conjugation length [24]. Therefore, dicyanostilbene-linked amphiphilic molecules are a promising turn-on fluorescent sensor for singlet oxygen ($^1O_2$) and hypochlorite (ClO^-). Thus, we designed triphenylphosphonium (TPP)-conjugated dicyanostilbene derivatives as amphiphilic molecules for turn-on detection of singlet oxygen ($^1O_2$) and hypochlorite (ClO^-). TPP lipophilic cations were linked to dicyanostilbene to construct more rigid nanostructures in aqueous solution (i.e., the lipophilic cations enhance the stability of organic nanostructures by preventing solvent hydration and providing temperature or pH resistance) [25]. Here, we report the self-assembly properties of mono- or di-TPP-conjugated dicyanostilbene derivatives (**1 and 2**) and their behaviors as selective turn-on fluorescence probes toward singlet oxygen ($^1O_2$) and hypochlorite (Figure 1). In Scheme 1, we represented the detection strategies of ROS with turn-on fluorescence of self-assembled probe **1**. The self-assembled probe **1** with turn-off changed to turn-on through morphological transformation from sphere to continuously networked fibrous structures.

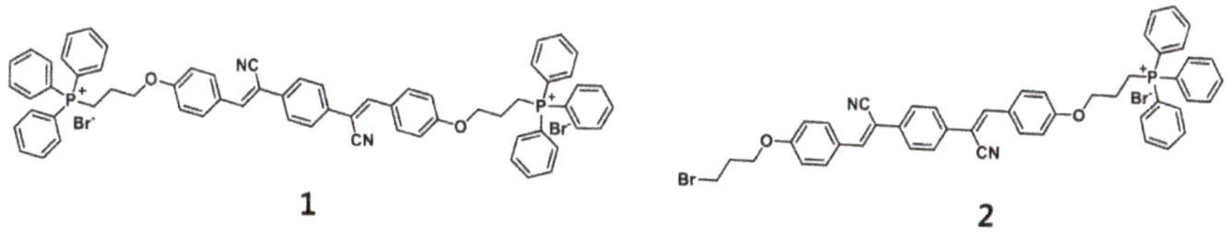

Figure 1. Chemical structures of probes **1** and **2**.

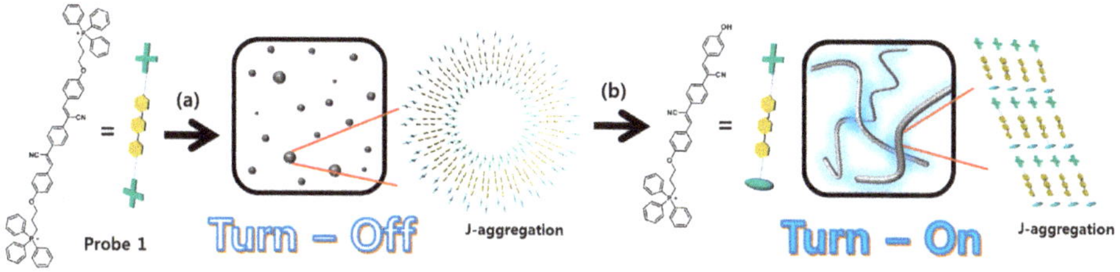

Scheme 1. Morphological change-based turn-on system for ROS detection: (**a**) formation of self-assembled spherical structure (gray sphere) and (**b**) morphological change into the fiber structure (gray ribbon) by ROS recognition.

2. Materials and Methods

2.1. Reagents and Instruments

All reagents were purchased from Sigma-Aldrich (Yongin, South Korea). The solvent was purchased from Samchun Pure Chemicals (Pyeongtaek, South Korea) and used with further purification. ^1H and ^{13}C NMR spectra were measured using a Bruker DRX 300 spectrometer (Bruker). Furthermore, the mass spectra were measured using a JEOL JMS-700 mass spectrometer (JEOL Ltd., Tokyo, Japan). In addition, a Thermo Evolution 600 UV-vis spectrophotometer (Thermo Fisher Scientific, Waltham, MA, USA) was used to obtain the absorption spectra in the solution, and the fluorescence spectra were recorded using a RF-5301PC spectrophotometer (Shimadzu Corp., Kyoto, Japan).

2.2. Synthesis of Compound 3

Compound **3** was synthesized using a previously reported method. Here 1,3-dibromopropane (2.5 mL, 24.57 mmol), *p*-hydroxybenzaldehyde (2.0 g, 16.38 mmol), and K_2CO_3 (3.39 g, 24.57 mmol) were dissolved in acetone (75 mL). The reaction mixture was heated to 50 °C for 12 h. Then, the reaction mixture was cooled to room temperature, filtered, and concentrated under reduced pressure. The crude product was purified using silica gel and eluted with ethyl acetate and hexane of compound **3**: ^1H NMR (300 MHz, DMSO-d_6) δ 9.88 (s, 1H), 7.90–7.86 (d, 2H), 7.17–7.14 (d, 2H), 4.23–4.19 (t, 2H), 3.70–3.66 (t, 2H), 2.32–2.24 (m, 2H); ^{13}C NMR (75 MHz, DMSO-d_6) δ 190.89, 163.89, 132.18, 130.38, 118.48, 114.99, 69.20, 65.91, 32.27, 29.78.

2.3. Synthesis of Compound 2

Compound **3** (1.71 g, 7.03 mmol) and p-xylene dicyanide (0.5 g, 3.2 mmol) were dissolved in ethanol (30 mL). Sodium methoxide (0.35 g, 6.4 mmol) was added, and the solution was heated at reflux under N_2 atmosphere. After 12, the reaction mixture was then cooled to room temperature, and the solid was filtered and washed with methanol: (yellow solid, 55% yield); ^1H NMR (300 MHz, DMSO-d_6), δ (ppm): 8.07 (s, 2H), 8.00–7.97 (d, 4H), 7.86 (s, 4 H), 7.17–7.14 (d, 4H), 4.22–4.18 (t, 4H), 3.70–3.68 (t, 4H), 2.34–2.25 (m, 4H); ^{13}C NMR (75 MHz, DMSO-d_6), δ (ppm): 160.3, 143.2, 134.4, 131.7, 126.0, 124.8, 118.5, 116.0, 105.1.

2.4. Synthesis of Probe 1

Compound **2** (0.5 g, 0.82 mmol) was dissolved in acetonitrile (100 mL) with triphenylphosphine (1.3 g, 4.92 mmol). The mixture was heated to 85 °C for 48 h. Acetonitrile was removed under vacuum, and the precipitated yellow solid was collected by recrystallization. Then, the product was purified by flash chromatography using dichloromethane: (yellow solid, 61% yield); ^1H NMR (300 MHz, DMSO-d_6), δ (ppm): 8.10 (s, 2H), 8.00–7.76 (m, 42H), 7.14–7.12 (t, 4H), 4.25–4.18 (t, 4H), 3.81–3.71 (t, 4H), 2.02(s, 4H); ^{13}C NMR (75 MHz, DMSO-d_6), δ (ppm): 160.6, 143.4, 165.5, 134.8, 134.2, 134.1, 131.8, 130.9, 130.7, 127.0, 126.6, 119.4, 118.7, 118.2, 115.6, 106.9; FT-IR (cm^{-1}): 2209, 1599, 1510, 1434, 1245, 1183, 1110; ESI-MS (m/z): 485.42 $[\mathbf{1} + H]^{2+}$, 1052.00 $[\mathbf{1} + Br]^+$.

2.5. Synthesis of Probe 2

Compound **2** (0.5 g, 0.82 mmol) was dissolved in acetonitrile (100 mL) with triphenylphosphine (0.43 g, 1.64 mmol). The mixture was heated to 85 °C for 48 h. Acetonitrile was removed under vacuum, and the precipitated yellow solid was collected by recrystallization. Then, the product was purified by flash chromatography using dichloromethane/methanol (10:1). (yellow solid, 22.3% yield); ^1H NMR (300 MHz, DMSO-d_6), δ (ppm): 8.09 (s, 2H), 8.01–7.76 (m, 23H), 7.14–7.11 (s, 4H), 4.25–4.19 (s, 4H), 3.84–3.67 (s, 4H), 2.23–2.21 (s, 2H), 2.05 (s, 2H); ^{13}C NMR (75 MHz, DMSO-d_6), δ (ppm): 161.5, 143.9,

136.1, 135.9, 130.8, 130.1, 126.2, 119.2, 117.6, 107.7, 68.2, 33.8, 31.2; FT-IR (cm^{-1}): 2214, 1594, 1508, 1434, 1508, 1434, 1259, 1176, 1109, 687; ESI-MS (m/z): 789.17 [**2** + H] $^+$.

2.6. Synthesis of **1-Ref**

1-Ref was synthesized by the reported method [S1].; ^{1}H NMR (300 MHz, DMSO-d_6), δ (ppm): 10.33 (s, 2H), 7.98 (s, 2H), 7.92–7.87 (d, 4H), 7.82 (s, 4H), 6.95 (s, 2H), 2); ^{13}C NMR (75 MHz, DMSO-d_6), δ (ppm): 160.4, 143.0, 133.4, 132.7, 126.5, 125.9, 117.4, 117.0, 105.2; FT-IR (cm^{-1}): 3200, 2222, 1609, 1592, 1514, 1440, 1300, 1173.

2.7. Fluorescence Spectroscopy

A 1-cm long cuvette was used in the fluorescence assay. The sample was excited at 367 nm, and the emission was collected from 500 nm. ROS detection experiments were performed three times. To detect the ROS by probe **1**, 1 mL of standard solution (pH = 7.4) was first added to the cuvette. Then, the hydroxyl radical was generated via Fenton reaction with different amounts of Fe^{2+} and H_2O_2 (Fe^{2+}/H_2O_2 = 1:10). After incubation with the probe for 15 min. For the selectivity experiment, hydroxyl radical ($^{\bullet}$OH) was generated via Fenton reaction (Fe^{2+}/H_2O_2 = 200 μM, 2000 μM). Superoxide anion ($O_2^{\bullet-}$) was derived from dissolved KO_2 (200 μM) in the DMSO solution. Hypochlorite anion (ClO^-) was provided by NaClO (200 μM). Nitric oxide (NO) and nitroxyl (HNO) were derived from a solution of S-nitroso-N-acetyl-DL-penicillamine and Angeli's salt, respectively. $^1O_2$ was generated by the reaction of H_2O_2 (200 μM) with NaClO (200 μM). Other species (10 equiv.) were prepared by dissolving in aqueous solutions at pH 7.4. All experiments were performed after incubation with the appropriate ROS/RNS for 10 min at room temperature.

2.8. Determination of Limit of Detection

The limit of detection (LOD) of probe **1** for ClO^- and $^1O_2$ was determined as 33 μM and 56 μM, respectively. The LOD was calculated using the following equation, where σ is the standard deviation of the blank measurements and s is the slope of the calibration plot.

$$LOD = 3 \times \sigma / s \tag{1}$$

3. Results and Discussion

3.1. Characterization of Self-Assembled Probes **1** and **2**

Probes **1** and **2** were synthesized by a reaction of bromine-modified dicyanostilbene with TPP in acetonitrile following a previously reported method. Compounds **1** and **2** were confirmed by ^{1}H and ^{13}C NMR, ESI-MS, and FT-IR spectroscopy (Figure S1). ^{1}H NMR data indicate that probes **1** and **2** demonstrated only a (Z)-form originating from the alkene peak at 8.04 in the dicyanostilbene moiety.

Since TPP-appended dicyanostilbene derivatives often demonstrate amphiphilic properties [26]. we observed the self-assembling behaviors in aqueous solution. In the aqueous solution (1% DMSO) at 25 °C, the UV-Vis absorption bands of **1** (25 μM) and **2** (6.25 μM) appeared at 360 nm and 371 nm, respectively. The UV-Vis absorption bands originated from π-π transitions of the dicyanostilbene moiety and shifted to a longer wavelength by increasing the temperatures of the aqueous solution (1% DMSO) (Figure 2). These red shifts were due to the formation of self-assembly via *J*-aggregation [27]. Note that the λ_{max} shift of probe **2** was smaller than that of probe **1**, which indicates that, compared to probe **2**, probe **1** formed more stable self-assembly in aqueous solution. The fluorescence spectra of probes **1** and **2** (excitation wavelength: **1** at 360 nm and **2** at 371 nm) were obtained by change of temperature (Figure S2). Weak fluorescence bands for probes **1** and **2** were observed at 512 nm due to the PET from fluorophore to TPP [28]. Due to the change from self-assembly to de-assembly, the fluorescence intensities decreased as the temperature increased. The fluorescence intensity of probe **1** decreased at 35–45 °C, whereas the fluorescence intensity of probe **2** decreased at 25–35 °C. We also measured the temperature-dependent ^{1}H NMR spectra of probes **1** and **2** to obtain the key factor in the formation of self-assembly in DMSO-d_6/D_2O (99/1, v/v%) (Figure S3). The tendency of the chemical shift of an alkene peak in probe **1** was similar to that of probe **2**; however, the shift of probe **1** was greater than that of probe **2**. In addition, the interaction between alkene groups may affect enhancement of self-assembly stability. However, compared to probe **2**, the stabilization enhancement of self-assembled probe **1** led to intrinsic fluorescence quenching due to the reduced distance between dicyanostilbene and TPP via *J*-aggregation, as well as the effect of TPP groups inducing PET from dicyanostilbene.

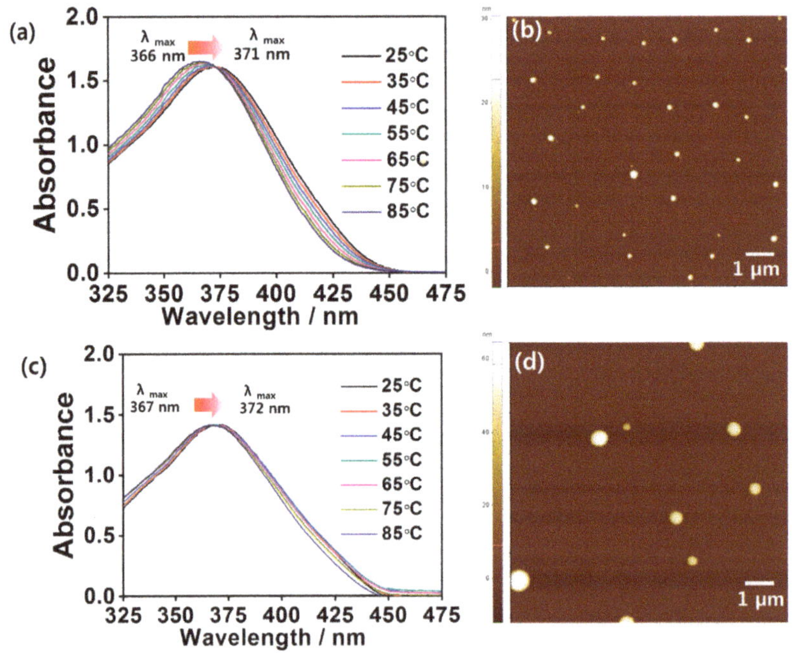

Figure 2. Temperature dependent UV–VIS spectra of probes (**a**) **1** (25 μM) and (**c**) **2** (6.25 μM) in DMSO/H_2O (1/99, v/v%). AFM images of probes (**b**) **1** (100 μM) and (**d**) **2** (100 μM).

We also observed morphologies of self-assembled **1** and **2** using atomic force microscopy (AFM) (Figures 2 and S9). The AFM image of probe **1** showed a spherical structure with a diameter of ca. 22–32 nm (Figures 2b and S9a). The spherical nanoparticle should form by intermolecular dipole-dipole interaction and π-π stacking [29]. Similarly, probe **2** showed a spherical structure with a diameter of ca. 53–63 nm (Figures 2d and S9b). The size difference of self-assembled spheres **1** and **2** was due to the TPP group and the binding strength of the π-π interaction between the alkene groups, as shown by the ^{1}H NMR data. Thus, we recognize that the size of the self-assembled spherical nanoparticles was determined by the strength of the intermolecular interactions. Based on UV-Vis, PL, and ^{1}H NMR experiments, we conclude that **1** and **2** formed self-assembled spherical nanoparticles by *J*-aggregation with dipole-dipole interaction between the alkene groups in the dicyanostilbene moiety. The stability of self-assembled **1** was greater than that of self-assembled probe **2** (i.e., the TPP substitution effect accompanied fluorescence turn-off). We inserted a table involving summarized characteristics of the probes **1** and **2** (Table S1).

3.2. *ROS-Sensing Ability of Self-Assembled Probes **1** and **2** in Aqueous Solution*

Self-assembled probes **1** and **2** with negligible fluorescence emission bands were used in aqueous solution to apply a turn-on fluorescence probe to highly ROS. The fluorescence spectral changes of self-assembled probe **1** were observed upon addition of several species related to ROS, such as singlet oxygen (10 equiv. of $^1O_2$) and other ROS (10 equiv. of O_2^-, H_2O_2, NO, TBHP, ClO^-, $^\bullet OH$, GSH, cysteine, histidine, K^+, and Na^+) in water at pH 7.4 (Figure 3). Upon treatment with 10 equiv. of $^1O_2$ and ClO^-, a marked strong green emission at 520 nm was observed in under 5 min, indicating that $^1O_2$ and ClO^- reacted with self-assembled **1** rapidly at room temperature. In addition, the fluorescence intensity of self-assembled **1** in the presence of $^1O_2$ and ClO^- was enhanced by 2.3 and 2.7 times, respectively, compared to self-assembled **1** in the absence of $^1O_2$ and ClO^-. In contrast, significant selective changes in the emission were not observed upon addition of O_2^-, H_2O_2, NO, TBHP, ClO^-, $^\bullet OH$, GSH, cysteine, histidine, K^+, and Na^+ (Figure 3), indicating that these ROS species, amino acids, and metal ions did not react to **1**. The large difference in the fluorescence images between $^1O_2$, ClO^-, and other ROS was observed after treatment of **1** (Figure 3). Their fluorescence enhancement of self-assembled **1** in the presence of $^1O_2$ or ClO^- indicated that the complex between **1** and $^1O_2$ or ClO^- hinders PET [18]. The non-emission of self-assembled **1** upon addition of other oxide species, such as O_2^- and NO, may be less reactive with 1 than that when $^1O_2$ or ClO^- are added [30]. The sensing ability of self-assembled **2** was also evaluated under the same conditions. Self-assembled **2** exhibited fluorescence enhancement upon addition of O_2^-, $^\bullet OH$, and ClO^-. A hydrolysis reaction of the bromide group in self-assembled **2** with O_2^- or $^\bullet OH$ did occur. The hydroxyl group in self-assembled **2** might induce fluorescence turn-off through the electron transfer mechanism [31]. Therefore, the fluorescence enhancement effect of self-assembled **2** was weaker than that of **1**. The slight turn-on effects of **2** in the presence of O_2^-, $^\bullet OH$, and ClO^- may be caused by the reaction of a –Br group with ROS (O_2^-, $^\bullet OH$, and ClO^-), as indicated by the IR data (Figure S4).

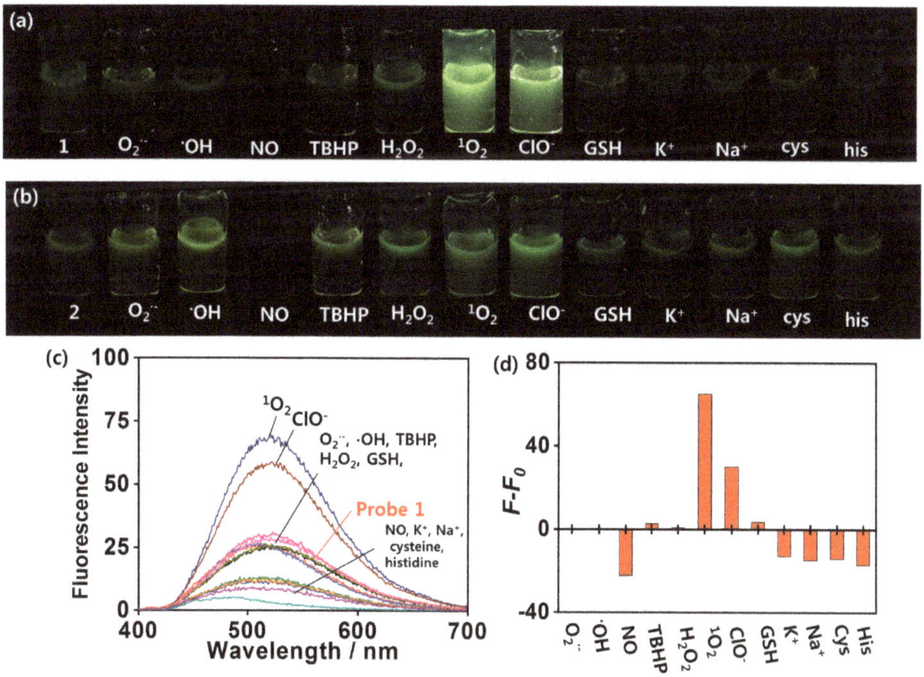

Figure 3. *Cont.*

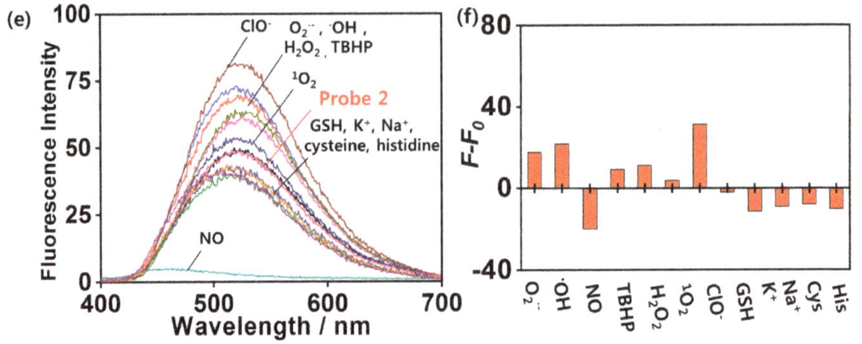

Figure 3. Photographs of (**a**) **1** (25 μM) and (**b**) **2** (6.25 μM) with ROS (10 equiv.), metal ions (10 equiv.), or amino acids (10 equiv.) in DMSO/H_2O (1/99, *v*/*v*%) under UV lamp. Fluorescence spectra (Excitation wavelength: 360 nm for probe **1** and 371 nm for probe **2**) of (**c**) **1** and (**e**) **2** with ROS (10 equiv.), metal ions (10 equiv.), or amino acids (10 equiv.) in DMSO/H_2O (1/99, *v*/*v*%). Fluorescence intensity of probes (**d**) **1** and (**f**) **2** in the presence of ROS, metal ions, and amino acids in DMSO/H_2O (1/99, *v*/*v*%).

To quantitatively investigate the reactivity and spectral changes of **1** (25 μM) upon addition of $^1O_2$ and ClO^-, fluorescence titrations were performed by adding $^1O_2$ and ClO^- (0–375 μM) in water (containing 1% DMSO) at room temperature (Figure 4). The fluorescence intensity of **1** at 520 nm, which originates from dicyanostilbene moiety, was enhanced drastically during the titration process. In addition, at less than **2** equiv. of $^1O_2$ and ClO^-, an excellent nonlinear correction between fluorescence intensity and the concentration of $^1O_2$ and ClO^- was obtained with R^2 = 0.9923 for $^1O_2$ and R^2 = 0.9633 for ClO^-, indicating that the ratio of fluorescence intensity at 520 nm was enhanced as a nonlinear function of $^1O_2$ and ClO^- concentration. This turn-on mechanism can be attributed to the cooperative effects of AIEE of dicyanostilbene and blocking of the photoinduced electron transfer process. The detection limits of self-assembled **1** for ClO^- and $^1O_2$ were 33 μM and 56 μM, respectively (Figure S5) [S2].

To further evaluate the utility of **1** as a selective fluorescence probe for $^1O_2$ and ClO^-, the competition-based fluorescence emission changes of self-assembled **1** upon addition of various biologically relevant species and ROS (i.e., ClO^-, $^1O_2$, $O_2{}^-$, H_2O_2, NO, TBHP, ClO^-, •OH, GSH, cysteine, histidine, K^+, and Na^+) were investigated in aqueous solution (Figure S6). In binary system, probe **1** showed a strong green emission with $^1O_2$ or ClO^- except for cysteine, histidine and NO. The emission change with cysteine was due to that the phenoxy oxygen of **1** may be interacted with -SH group in cysteine [32,33]. In addition, histidine molecules were due to their bioactive properties inducing the reaction of histidine with $^1O_2$ or ClO^- prior to probe **1** [33]. Low turn-on emission of probe **1** in presence of K^+ and Na^+ ions can be caused by the charge interaction of cation with $^1O_2$ or ClO^- prior to react with probe **1**. As expected, the fluorescence intensities of self-assembled **1** in the presence of $^1O_2$ and ClO^- were unchanged by treatment of other species, such as $O_2{}^-$, H_2O_2, NO, TBHP, •OH, GSH, K^+, and Na^+, indicating that self-assembled **1** is a new selective turn-on fluorescence probe of $^1O_2$ and ClO^- in a mixture of other species.

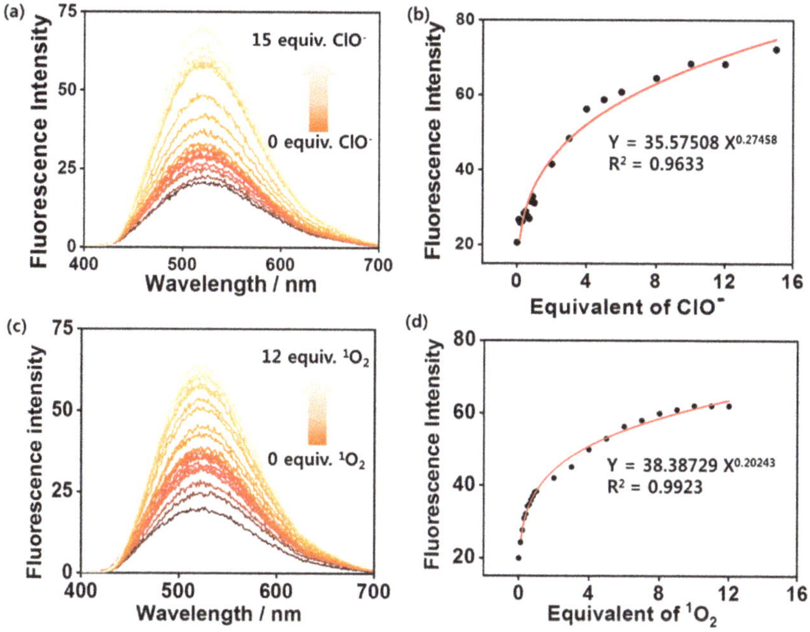

Figure 4. Fluorescence spectra (Excitation wavelength: 360 nm for probe **1**) of **1** (25 μM) in the presence of various concentrations of (**a**) ClO^- (0–375 μM) and (**c**) $^1O_2$ (0–300 μM) in DMSO/H_2O (1/99, *v*/*v*%). Plot for fluorescence intensity of **1** upon addition of (**b**) ClO^- and (**d**) $^1O_2$.

3.3. ROS-Mediated Fluorescence Turn-On Mechanism of Self-Assembled Probe 1 in Aqueous Solution

The mechanism of the reaction between self-assembled **1** and $^1O_2$ or ClO^- was further studied. First, the bonding cleavage of **1** upon addition of $^1O_2$ or ClO^- was confirmed at the molecular level. The mechanism of the reaction between self-assembled **1** and $^1O_2$ or ClO^- was further studied. First, the bonding cleavage of **1** upon addition of $^1O_2$ or ClO^- was confirmed at the molecular level. The shape of the UV-Vis spectra of self-assembled probe **1** treated with $^1O_2$ or ClO^- was almost same to that without $^1O_2$ or ClO^- (Figure S7). On the other hand, the molecular structure of dicyanostilbene was conserved when $^1O_2$ or ClO^- was added to self-assembled **1**. We observed FAB-Mass and IR spectral changes of **1** with $^1O_2$ or ClO^- (Figure S7). After treatment with $^1O_2$ or ClO^-, a mass value of 667.3 was obtained. This value indicates that the oxygen atom adjacent at dicyanostilbene reacted with $^1O_2$ or ClO^- and then formed a phenol moiety (Figure S7). Furthermore, the IR spectra of **1** upon addition of $^1O_2$ or ClO^- showed new peaks at 900–1100 cm^{-1} (OH bending vibration), and peaks in the range of 3100–3700 cm^{-1} were widened due to the formation of the phenolic OH group in **1** (Figure S7). In addition, by comparing the IR data (1000–2400 cm^{-1}) of self-assembled **1** with or without analyte, we confirmed that the structure of the dicyanostilbene moiety was conserved while self-assembled **1** reacted to $^1O_2$ or ClO^-. A dicyanostilbene derivative possessing OH group (**1-Ref**) was synthesized to assign OH groups on IR and NMR data (Figures S7A and S8). In Figures S7 and S8, the OH groups the product obtained from probe **1** in the presence of $^1O_2$ or ClO^- correspond with the OH peaks compound **1-Ref**. However, the product obtained from probe **1** with $^1O_2$ or ClO^- was produced one −OH as shown in Figure S8, but not two −OH groups. We measured zeta potential of probe **1** and probe **1** treated with $^1O_2$ or ClO^- to further confirm about its structure change (Figure S10). The zeta potential of the probe **1** was determined to be 32.46 mV; however, the zeta potential of the probe **1** treated with $^1O_2$ or ClO^- was 21.43 mV and 23.16 mV respectively; this was due to the formation of

OH group. These results strongly indicated that the OH group originated from the reaction of probe **1** with $^1O_2$ or ClO^- had been successfully formed.

Furthermore, the morphological change of self-assembled **1** was observed by treatment with $^1O_2$ or ClO^- by AFM. After treatment with $^1O_2$ or ClO^-, the spherical nanoparticle formed via self-assembly of **1** changed to a fiber structure (Figures 5 and S9c).

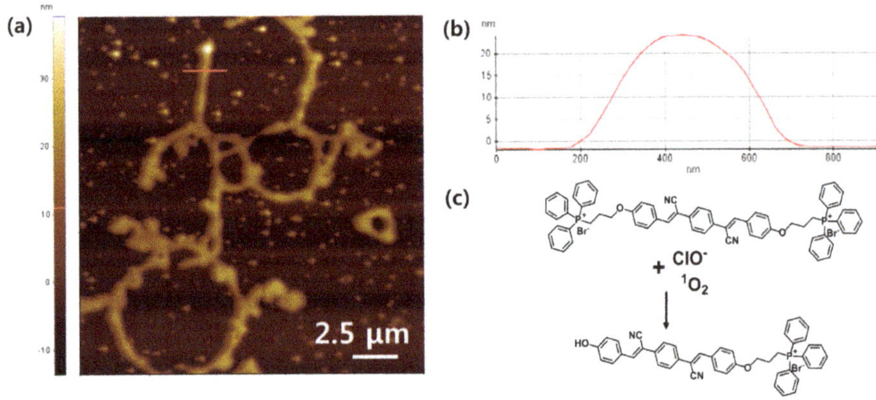

Figure 5. (**a**) AFM image of **1** (100 µM) in DMSO/H_2O (1/99, *v*/*v*%) after adding ClO^- or $^1O_2$. (**b**) Height and width of nanofiber (**a**). (**c**) Mechanism of **1** after adding ClO^- or $^1O_2$. (ClO^--treated probe **1** and $^1O_2$-treated probe **1** were shown similar morphologies).

Therefore, we propose a schematic illustration of the reaction-based morphological change of self-assembled **1** during the reaction that progressed upon adding $^1O_2$ or ClO^- (Figure 5). To the best of our knowledge, the selectivity of probe **1** for $^1O_2$ could be described by two-step reactions. First, an aliphatic carbon adjacent to the oxygen atom was attacked by $^1O_2$ [34]. Then, water reacted with the $^1O_2$-derived activated group to form a hydroxyl end group in the dicyanostilbene moiety. Due to the structural stabilization effect of the fully-conjugated dicyanostilbene moiety, alkene groups in dicyanostilbene may be less reactive than the carbon atoms adjacent to the oxygen atom of probe **1**.

4. Conclusions

We have synthesized novel triphenylphosphonium-strapped dicyanostilbene derivatives (probes **1** and **2**) and characterized their self-assembly properties by using spectroscopic analysis. Owing to the TPP substitution effects, probe **1** were formed more stable self-assembly than that of probe **2**. The spherical structure of self-assembled probe **1** were shown selectively turn-on fluorescence upon addition of $^1O_2$ or ClO^-. Moreover, the morphology of self-assembled probe **1** was changed from sphere to fibrous structures in the presence of $^1O_2$ or ClO^-. We also proved that the generation of OH group-substituted TPP-dicyanostilbene caused by the reaction between oxygen atoms adjacent dicyanostilbene of probe **1** and $^1O_2$ or ClO^-. Thus, we could propose continuous fibrous-mediated AIEE effect-based turn-on sensing mechanism of probe **1** for $^1O_2$ or ClO^- by using molecular and nanometer level analysis. Based on our results, it is expected that a nanostructure change-derived turn-on sensing strategy for $^1O_2$ or ClO^- may offer a new approach to develop methods in biological and environmental subjects.

Supplementary Materials: The supplementary materials are available online at http://www.mdpi.com/2079-4991/8/12/1034/s1.

Author Contributions: Conceptualization, W.C. and N.Y.L.; Data curation, W.C.; Formal analysis, W.C. and N.Y.L.; Investigation, W.C., N.Y.L. and H.C.; Methodology, H.C.; Project administration, H.C. and M.L.S.; Supervision, M.L.S., J.A. and J.H.J.; Writing-original draft, J.A.; Writing-review & editing, J.A. and J.H.J.

Funding: This research was supported by the NRF (2018R1A2B2003637 and 2017R1A4A1014595) supported by the Ministry of Education, Science and Technology, Korea. In addition, this work was partially supported by a grant from the Next-Generation BioGreen 21 Program (SSAC, Grant no. PJ013186052018), RuralDevelopment Administration, Korea.

Conflicts of Interest: The authors declare no conflict of interest.

References

1. Zhao, X.; Zheng, H.; Qu, D.; Jiang, H.; Fan, W.; Sun, Y.; Xu, Y. A supramolecular approach towards strong and tough polymer nanocomposite fibers. *RSC Adv.* **2018**, *8*, 10361–10366. [CrossRef]
2. Cherumukkil, S.; Vedhanarayanan, B.; Das, G.; Praveen, V.K.; Ajayaghosh, A. Self-assembly of Bodipy-derived extended π-Systems. *Bull. Chem. Soc. Jpn.* **2018**, *91*, 100–120. [CrossRef]
3. Shimizu, T. Self-assembly of discrete organic nanotubes. *Bull. Chem. Soc. Jpn.* **2018**, *91*, 623–668. [CrossRef]
4. Ariga, K.; Nishikawa, M.; Mori, T.; Takeya, J.; Shrestha, L.K.; Hill, J.P. Self-assembly as a key player for materials nanoarchitectonics. *Sci. Technol. Adv. Mater.* **2018**. [CrossRef]
5. Li, S.; Tian, T.; Zhang, T.; Cai, X.; Lin, Y. Advances in biological applications of self-assembled DNA tetrahedral nanostructures. *Mater. Today* **2018**. [CrossRef]
6. Hu, Z.-T.; Chen, Z.; Goei, R.; Wu, W.; Lim, T.-T. Magnetically recyclable Bi/Fe-based hierarchical nanostructures via self-assembly for environmental decontamination. *Nanoscale* **2016**, *8*, 12736–12746. [CrossRef] [PubMed]
7. Wang, H.; Hu, L.; Du, W.; Tian, X.; Hu, Z.; Zhang, Q.; Zhou, H.; Wu, J.; Uvdal, K.; Tian, Y. Mitochondria-targeted iridium (III) complexes as two-photon fluorogenic probes of cysteine/homocysteine. *Sens. Actuators B* **2018**, *255*, 408–415. [CrossRef]
8. Wong, J.K.H.; Todd, M.H.; Rutledge, P.J. Recent advances in macrocyclic fluorescent probes for ion sensing. *Molecules* **2017**, *22*, 200. [CrossRef] [PubMed]
9. Mandal, K.; Jana, D.; Ghorai, B.K.; Jana, N.R. Functionalized chitosan with self-assembly induced and subcellular localization-dependent fluorescence 'switch on' property. *New J. Chem.* **2018**, *42*, 5774–5784. [CrossRef]
10. Shi, J.; Wu, Y.; Tong, B.; Zhi, J.; Dong, Y. Tunable fluorescence upon aggregation: Photophysical properties of cationic conjugated polyelectrolytes containing AIE and ACQ units and their use in the dual-channel quantification of heparin. *Sens. Actuators B* **2014**, *197*, 334–341. [CrossRef]
11. Chen, X.; Tian, X.; Shin, I.; Yoon, J. Fluorescent and luminescent probes for detection of reactive oxygen and nitrogen species. *Chem. Soc. Rev.* **2011**, *40*, 4783–4804. [PubMed]
12. Jin, L.; Xu, M.; Jiang, H.; Wang, W.; Wang, Q. A simple fluorescein derived colorimetric and fluorescent 'off-on' sensor for the detection of hypochlorite. *Anal. Methods* **2018**, *10*, 4562–4569. [CrossRef]
13. Li, K.; Hou, J.-T.; Yang, J.; Yu, X.-Q. A tumor-specific and mitochondria-targeted fluorescent probe for real-time sensing of hypochlorite in living cells. *Chem. Commun.* **2017**, *53*, 5539–5541. [CrossRef] [PubMed]
14. Pavelescu, L.A.; Iordache, M.-M.; Savopol, T.; Kovacs, E.; Moisescu, M.G. A new technique for evaluating reactive oxygen species generation. *Biointerface Res. Appl. Chem.* **2015**, *5*, 1003–1006.
15. Jiao, X.; Li, Y.; Niu, J.; Xie, X.; Wang, X.; Tang, B. Small-Molecule Fluorescent Probes for Imaging and Detection of Reactive Oxygen, Nitrogen, and Sulfur Species in Biological Systems. *Anal. Chem.* **2018**, *90*, 533–555. [CrossRef]
16. Chang, C.; Wang, F.; Qiang, J.; Zhang, Z.; Chen, Y.; Zhang, W.; Wang, Y.; Chen, X. Benzothiazole-based fluorescent sensor for hypochlorite detection and its application for biological imaging. *Sens. Actuators B* **2017**, *243*, 22–28. [CrossRef]
17. Yu, H.; Liu, X.; Wu, Q.; Li, Q.; Wang, S.; Guo, Y. A new rhodamine-based fluorescent probe for the detection of singlet oxygen. *Chem. Lett.* **2015**, *44*, 244–246. [CrossRef]
18. Liu, H.-W.; Xu, S.; Wang, P.; Hu, X.-X.; Zhang, J.; Yuan, L.; Zhang, X.-B.; Tan, W. An efficient two-photon fluorescent probe for monitoring mitochondrial singlet oxygen in tissues during photodynamic therapy. *Chem. Commun.* **2016**, *52*, 12330–12333. [CrossRef]
19. You, Y.; Cho, E.J.; Kwon, H.; Hwang, J.; Lee, S.E. A singlet oxygen photosensitizer enables photoluminescent monitoring of singlet oxygen doses. *Chem. Commun.* **2016**, *52*, 780–783. [CrossRef]

20. Mulay, S.V.; Choi, M.; Jang, Y.J.; Kim, Y.; Jon, S.; Churchill, D.G. Enhanced Fluorescence Turn-on Imaging of Hypochlorous Acid in Living Immune and Cancer Cells. *Chem. Eur. J.* **2016**, *22*, 9642–9648. [CrossRef]
21. Wang, L.; Yang, L.; Zhu, L.; Cao, D.; Li, L. Synthesis, characterization and fluorescence "turn-on" detection of BSA based on the cationic poly(diketopyrrolopyrrole-co-ethynylfluorene) through deaggregating process. *Sens. Actuators B* **2016**, *231*, 733–743. [CrossRef]
22. Alifu, N.; Dong, X.; Li, D.; Sun, X.; Zebibula, A.; Zhang, D.; Zhang, G.; Qian, J. Aggregation-induced emission nanoparticles as photosensitizer for two-photon photodynamic therapy. *Mater. Chem. Front.* **2017**, *1*, 1746–1753. [CrossRef]
23. En, Y.; Zhang, R.; Yan, C.; Wang, T.; Cheng, H.; Cheng, X. Self-assembly, AIEE and mechanochromic properties of amphiphilic α-cyanostilbene derivatives. *Tetrahedron* **2017**, *73*, 5253–5259.
24. Luo, C.; Liu, Y.; Zhang, Q.; Cai, X. Hyperbranched conjugated polymers containing 1,3-butadiene units: Metal-free catalyzed synthesis and selective chemosensors for Fe^{3+} ions. *RSC Adv.* **2017**, *7*, 12269–12276. [CrossRef]
25. Xiao, H.; Li, J.; Zhao, J.; Yin, G.; Quan, Y.; Wang, J.; Wang, R. A colorimetric and ratiometric fluorescent probe for ClO- targeting in mitochondria and its application in vivo. *J. Mater. Chem. B* **2015**, *3*, 1633–1638. [CrossRef]
26. Kim, K.Y.; Jin, H.; Park, J.; Jung, S.H.; Lee, J.H.; Park, H.; Kim, S.K.; Bae, J.; Jung, J.H. Mitochondria-targeting self-assembled nanoparticles derived from triphenylphosphonium-conjugated cyanostilbene enable site-specific imaging and anticancer drug delivery. *Nano Res.* **2018**, *11*, 1082–1098. [CrossRef]
27. Tay-Agbozo, S.; Street, S.; Kispert, L.D. The carotenoid bixin: Optical studies of aggregation in polar/water solvents. *J. Photochem. Photobiol. A* **2018**, *362*, 31–39. [CrossRef]
28. Yuan, H.; Cho, H.; Chen, H.H.; Panagia, M.; Sosnovik, D.E.; Josephson, L. Fluorescent and radiolabeled triphenylphosphonium probes for imaging mitochondria. *Chem. Commun.* **2013**, *49*, 10361–10363. [CrossRef]
29. Van Herrikhuyzen, J.; Willems, R.; George, S.J.; Flipse, C.; Gielen, J.C.; Christianen, P.C.M.; Schenning, A.P.H.J.; Meskers, S.C.J. Atomic Force Microscopy Nanomanipulation of Shape Persistent, Spherical, Self-Assembled Aggregates of Gold Nanoparticles. *ACS Nano* **2010**, *4*, 6501–6508. [CrossRef]
30. Hu, J.J.; Wong, N.-K.; Lu, M.-Y.; Chen, X.; Ye, S.; Zhao, A.Q.; Gao, P.; Kao, R.Y.-T.; Shen, J.; Yang, D. HKOCl-3: A fluorescent hypochlorous acid probe for live-cell and in vivo imaging and quantitative application in flow cytometry and a 96-well microplate assay. *Chem. Sci.* **2016**, *7*, 2094–2099. [CrossRef]
31. Zhang, H.; Liu, J.; Liu, C.; Yu, P.; Sun, M.; Yan, X.; Guo, J.-P.; Guo, W. Imaging lysosomal highly reactive oxygen species and lighting up cancer cells and tumors enabled by a Si-rhodamine-based near-infrared fluorescent probe. *Biomaterials* **2017**, *133*, 60–69. [CrossRef] [PubMed]
32. Njeri, C.W.; Ellis, H.R. Shifting redox states of the iron center partitions CDO between crosslink formation or cysteine oxidation. *Arch. Biochem. Biophys.* **2014**, *558*, 61–69. [CrossRef] [PubMed]
33. Hou, J.; Zhang, F.; Yan, X.; Wang, L.; Yan, J.; Ding, H.; Ding, L. Sensitive detection of biothiols and histidine based on the recovered fluorescence of the carbon quantum dots-Hg (II) system. *Anal. Chim. Acta* **2015**, *859*, 72–78. [CrossRef] [PubMed]
34. Sagadevan, A.; Hwang, K.C.; Su, M.-D. Singlet oxygen-mediated selective C–H bond hydroperoxidation of ethereal hydrocarbons. *Nat. Commun.* **2017**, *8*, 1812. [CrossRef] [PubMed]

nanomaterials

Article

A Modular Coassembly Approach to All-In-One Multifunctional Nanoplatform for Synergistic Codelivery of Doxorubicin and Curcumin

Muyang Yang [1,2,3], Lixia Yu [1,2], Ruiwei Guo [1,2], Anjie Dong [1,2], Cunguo Lin [3] and Jianhua Zhang [1,4,*]

1 Department of Polymer Science and Technology, Key Laboratory of Systems Bioengineering (Ministry of Education), School of Chemical Engineering and Technology, Tianjin University, Tianjin 300072, China; yangmuyang@tju.edu.cn (M.Y.); lixiayu@tju.edu.cn (L.Y.); rwguo@tju.edu.cn (R.G.); ajdong@tju.edu.cn (A.D.)

2 Collaborative Innovation Center of Chemical Science and Engineering (Tianjin), Tianjin 300072, China

3 State Key Laboratory for Marine Corrosion and Protection, Luoyang Ship Material Research Institute (LSMRI), Qingdao 266101, China; lincg@sunrui.net

4 Tianjin Key Laboratory of Membrane Science and Desalination Technology, Tianjin University, Tianjin 300072, China

* Correspondence: jhuazhang@tju.edu.cn; Tel: +86-22-2470-2364

Received: 21 February 2018; Accepted: 13 March 2018; Published: 15 March 2018

Abstract: Synergistic combination therapy by integrating chemotherapeutics and chemosensitizers into nanoparticles has demonstrated great potential to reduce side effects, overcome multidrug resistance (MDR), and thus improve therapeutic efficacy. However, with regard to the nanocarriers for multidrug codelivery, it remains a strong challenge to maintain design simplicity, while incorporating the desirable multifunctionalities, such as coloaded high payloads, targeted delivery, hemodynamic stability, and also to ensure low drug leakage before reaching the tumor site, but simultaneously the corelease of drugs in the same cancer cell. Herein, we developed a facile modular coassembly approach to construct an all-in-one multifunctional multidrug delivery system for the synergistic codelivery of doxorubicin (DOX, chemotherapeutic agent) and curcumin (CUR, MDR modulator). The acid-cleavable PEGylated polymeric prodrug (DOX-*h*-PCEC), tumor cell-specific targeting peptide (CRGDK-PEG-PCL), and natural chemosensitizer (CUR) were ratiometrically assembled into in one single nanocarrier (CUR/DOX-*h*-PCEC@CRGDK NPs). The resulting CUR/DOX-*h*-PCEC@CRGDK NPs exhibited several desirable characteristics, such as efficient and ratiometric drug loading, high hemodynamic stability and low drug leakage, tumor intracellular acid-triggered cleavage, and subsequent intracellular simultaneous drug corelease, which are expected to maximize a synergistic effect of chemotherapy and chemosensitization. Collectively, the multifunctional nanocarrier is feasible for the creation of a robust nanoplatform for targeted multidrug codelivery and efficient MDR modulation.

Keywords: modular coassemble; synergistic codelivery; polymeric prodrug; stimulisensitive release; biocompatibility

1. Introduction

Cancer is a leading death cause worldwide, and thus an urgent need constantly exists to improve its therapeutic outcomes [1]. Currently, chemotherapy with the advantages of minimal invasion and convenient administration is still regarded to be one of the most appropriate strategies to treat cancer patients [2,3]. However, the chemotherapeutic efficacy, especially the single-agent chemotherapy, is still far from satisfactory. In addition of the complicated pathogenesis and diversified pathologies of cancers, the major reason for treatment failure can be ascribed to the intrinsic drawbacks of chemotherapeutics [3–7]. For example, most anti-cancer drugs often suffer from poor water solubility and low stability, causing low bioavailability and poor pharmacokinetics. In addition, chemotherapeutic agents are unable to discriminate between cancerous and normal cells, resulting in non-specific distribution, and thus severe systemic toxicity. Moreover, the development of multidrug resistance (MDR) that is associated with the overexpression of P-glycoproteins (P-gp) poses a tremendous challenge to effective cancer therapy [5,8]. Therefore, numerous efforts are still required to address these challenges and improve chemotherapeutic efficacy.

During the past decades, in view of the complexity of cancer and thus the limited clinical efficacy of monotherapy, the administration of two or more drugs with complementary anticancer mechanisms, i.e., combination chemotherapy, has become increasingly important to enhance anticancer efficacy [9–13]. In particular, the nanoparticulate combination therapy, i.e., codelivery of multiple drugs by nanoparticles (NPs), merged the beneficial features of both nanoformulations and combination therapy, which have demonstrated to be an effective strategy to treat cancer [12,14–17]. When compared with the conventional cocktail chemotherapy, the codelivery of multiple drugs in one NP system can realize the definitive delivery of a correct ratio of each drug and enable the controlled release in target site, prolong drug circulation half-life, accumulate at the tumors through enhanced permeation and retention (EPR) effect, and thus optimize the pharmacokinetics and biodistribution of drugs, achieving a more significant synergistic anticancer effect [14–17]. For example, the concurrent delivery of doxorubicin (DOX, one of the most active chemotherapeutics for cancer) and curcumin (CUR, a natural chemosensitizer with distinguishing abilities to inhibit P-gp overexpression and nuclear transcription factor NF-κB activation that are both closely linked to MDR) in one nanocarrier has been widely explored as an effective way to improve DOX treatment efficiency [18–21]. A variety of nanoparticulate delivery systems, such as liposomes or lipid NPs [21–24], polymeric micelles [25–27], prodrug NPs [28–30], as well as inorganic NPs [31–33], have been developed as codelivery vesicles for DOX/CUR. The codelivery of DOX/CUR with nanotechnology has achieved significant success in improving the therapeutic efficacy of DOX [18–21]. However, the nanoparticle-assisted combination therapies are still far from fulfilling their potential and problems still exist in many aspects [14–16,18,34].

Firstly, due to the different physicochemical profiles and distinct solubility characteristics, it remains a great challenge to coencapsulate high payloads of chemotherapeutic agents and chemosensitizers into a single NP to unify their pharmacokinetics and biodistribution, and also to ensure no drug leakage and no rapid clearance into reticuloendothelial organs during in vivo transportation [14,15,18]. For example, the liposomes and polymeric micelles for multidrug codelivery often suffer from their low encapsulation capacity, instability, and premature drug leakage, meanwhile the applications of inorganic NPs frequently lead to their low biocompatibility and clearance of the immune system [3,4,35]. Another major challenge of the multidrug codelivery is the ability to improve the specificity to the disease site. Currently, most of the existing nanocarriers for combination therapy lack selectivity and targeting ability. Thus, their delivery into cancer tissues mostly relies on the EPR effect by the leaky tumor vasculatures varying among patients and their tumor types [36,37]. The active targeting, i.e., ligand-mediated targeting, involves using affinity ligands on the NPs surface for specific retention and uptake by tumor cells, which has been proven to be a promising complementary strategy for EPR in order to further improve the efficiency of cancer nanomedicines [36,38–40]. Apparently, the actively targeted NPs for combination drug delivery, without tedious steps for preparing to assure future scale-up, are in urgent need for enhanced therapeutic activity and reduced damage to healthy tissue. Finally, the nanocarriers should be able to effectively corelease the encapsulated chemotherapeutic drug and chemosensitizer into the same cancer cell at the same time after systemic drug administration, as the simultaneous intracellular corelease is the prerequisite for generation of synergistic responses [7,14,18]. Summarily, due to the different physicochemical and pharmacokinetic profiles, the synergistic codelivery of a chemotherapeutic agent and chemosensitizer will necessarily require further development of multifunctional nanocarriers with the capability of efficient encapsulation, targeted delivery, and intracellular controlled corelease.

In this study, a modular coassembly of acid-cleavable PEGylated polymeric prodrug, tumor cell-specific targeting peptide and natural chemosensitizer was designed and developed to construct multifunctional multidrug delivery system, as shown in Scheme 1. The chemotherapeutic agent DOX was conjugated to both ends of a biocompatible and biodegradable poly(ε-caprolactone)-*b*-poly(ethylene glycol)-*b*-poly(ε-caprolactone) (PCEC) via acid-cleavable hydrazone linkages, generating acid-cleavable PEGylated polymeric prodrug (DOX-*h*-PCEC) as drug vector model. To further introduce a tumor targeting capacity, cell-penetrating peptide (Cys−Arg−Gly−Asp−Lys, CRGDK) decorated poly(ethylene glycol)-*b*-poly(ε-caprolactone) (CRGDK-PEG-PCL, as active targeting model) was prepared via a maleimide-thiol reaction between CRGDK and maleimide-terminated PEG-PCL. CRGDK-mediated targeting can actively recognize the corresponding receptors (neuropilin-1 receptor, a transmembrane receptor glycoprotein overexpressed on the surface of a wide variety of tumor cells), which has been widely demonstrated to be able to increase the affinity for tumor cells and facilitate the drug-loaded NPs to efficiently enter the cells through ligand-receptor mediated endocytosis [41–45]. The DOX-*h*-PCEC and CRGDK-PEG-PCL are amphiphilic, and thus can coassemble in water, together with hydrophobic natural chemosensitizer (CUR, as MDR modulator model), forming a DOX/CUR coencapsulated, tumor-targeted, and intracellular acid-responsive nanoparticulate combination therapy system (CUR/DOX-*h*-PCEC@CRGDK NPs). With such unique structure, composition, and fabrication method, the CUR/DOX-*h*-PCEC@CRGDK NPs are expected to possess several desired functions: (1) high and tunable loading capacity derived from the PCL hydrophobic core and DOX hydrophobic phase, allowing for hydrophobic interaction and strong intermolecular π–π stacking action between DOX and CUR; (2) high stability and low leakage at physiological pH due to the combination of chemical conjugation and physical interactions; (3) cancer targeted delivery by the EPR effect and the active targeting of CRGDK; (4) stealth-shielding and long circulation from PEG shell; and, (5) intracellular simultaneous corelease of DOX and CUR through acid-triggered degradation of prodrug and subsequent disassembly of NPs to maximize a synergistic effect. Notably, the modular coassembly strategy to these all-in-one multifunctional NPs for DOX/CUR codelivery is cost-effective, as the complicated reactions and tedious synthesis processes for incorporating the desired multifunctionalities were minimized. Collectively, the design presented here provides a facile and robust nanoplatform for the targeted multidrug codelivery to maximize synergistic effects, holding great promises for the combinatory cancer therapy.

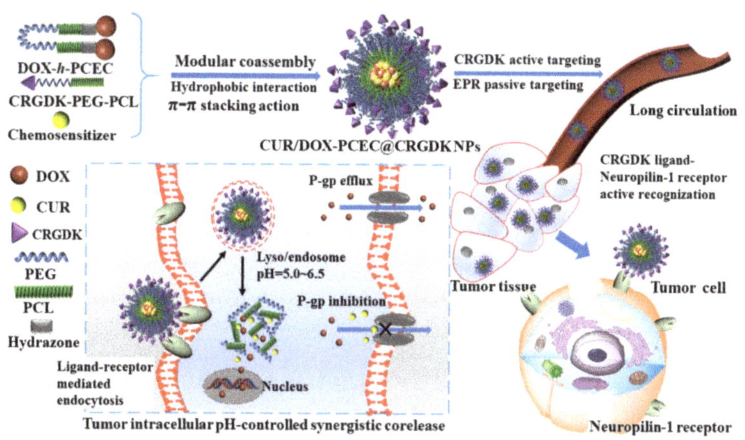

Scheme 1. Schematic illustrations of the modular coassembly to construct an all-in-one multifunctional multidrug delivery system (curcumin (CUR)/acid-cleavable PEGylated polymeric prodrug (DOX-*h*-PCEC) @Cys−Arg−Gly−Asp−Lys (CRGDK) nanoparticles (NPs)) with the capability of efficient encapsulation, targeted delivery, and intracellular controlled corelease.

2. Materials and Methods

2.1. Materials

Poly (ethylene glycol) (PEG, M_n = 1500 Da), ε-caprolactone (CL, 99%) and tert-butyl carbazate (Boc-$NHNH_2$, 99%) were purchased from Aladdin Industrial Co., Ltd. (Shanghai, China). Stannous octoate ($Sn(Oct)_2$, 99%) was obtained from Sigma-Aldrich China (Shanghai, China). 4-Nitrophenyl chloroformate (NPC, 99%), trimethylamine (TEA, 99%), and trifluoroacetic acid (TFA, 99%) was purchased from J&K chemical Co. Ltd. (Beijing, China). Doxorubicin (DOX, 99%) and curcumin (CUR, 99%) were purchased from Dalian Meilun biotechnology Co., Ltd. (Dalian, China). Maleimide-poly(ethylene glycol)- hydroxy (Mal-PEG-OH) with the molecular weight of 2000 was purchased from JenKem Technology Co., Ltd. (Beijing, China). Cys–Arg–Gly–Asp–Lys (CRGDK, 99%) peptide was provided by GL Biochem. Ltd. (Shanghai, China). Dichloromethane (DCM, 99%), dimethyl sulfoxide (DMSO, 99%), tetrahydrofuran (THF, 99%), and other reagents were commercially available from Damao Chemical Co., Ltd. (Tianjin, China). *N*,*N*-dimethylformamide (DMF, 99%) for HPLC Analysis was purchased form Merck (Darmstadt, Germany).

The Neuropilin-1 overexpressed human umbilical vein endothelial cells (HUVEC) and Adriamycin resistant breast cancer cells (MCF-7/ADR) were obtained from Procell life science and technology Co., Ltd. (Wuhan, China).

2.2. Synthesis and Characterization of DOX-h-PCEC and CRGDK-PEG-PCL

The hydrazone-linked DOX polymeric prodrug DOX-hydrazone-poly(ε-caprolactone)-*b*-poly(ethyleneglycol)-*b*-poly(ε-caprolactone)-hydrazone-DOX (DOX-*h*-PCEC) was prepared in three steps, according to a modified procedure [46], including (I) ring-opening polymerization of CL with HO-PEG-OH to prepare PCEC based on our group's previous work [47]; (II) introduction of the hydrazide group to both ends of PCEC; and, (III) conjugation of DOX to both ends of PCEC. After preparing PCEC (PCL_{17}-PEG_{34}-PCL_{17}, $M_n \approx 5.3$ kDa), the hydrazide-functionalized PCEC was prepared, as follows. Firstly, PCEC (1.05 g, 0.2 mmol) and TEA (83.4 μL, 0.6 mmol) were dissolved in flask with appropriate THF (10 mL). After incubation in the ice-water bath, NPC (96.7 mg, 0.48 mmol) dissolved in THF (5 mL) was dropwise added into the flask and the reaction lasted for 8 h. The resultant salts were removed by filtration and then the NPC-activated PCEC was precipitated from cold ether and dried in vacuum. Subsequently, the NPC-activated PCEC (about 1.13 g, 0.2 mmol) was dissolved in DMF and reacted with Boc-$NHNH_2$ (66.1 mg, 0.5 mmol) for 12 h. The resultant mixture was treated by TFA (2.5 mL) for 4 h at room temperature to remove -$COOC(CH_3)_3$ to obtain hydrazide-functionalized PCEC. After removing DMF by dialysis in water using a dialysis membrane with a molecular weight cutoff (MWCO) of 3.5 kDa, the hydrazide-functionalized PCEC was obtained by lyophilization. Finally, hydrazide-functionalized PCEC (1.08 g, 0.2 mmol) and excess DOX (0.272 g, 0.5 mmol) were dissolved in DMSO (20 mL) and reacted for 24 h at 30 °C in the presence of acetic acid as catalyst (0.5 mL). Unreacted DOX were removed using dialysis for 24 h against DMSO. Then, DMSO was removed by dialyzing for 24 h against PBS buffer (pH 7.4). After lyophilization, DOX-*h*-PCEC was obtained at a yield of 89%. All of the reactions were performed in a dark room.

CRGDK-conjugated poly (ethylene glycol)-*b*-poly (ε-caprolactone) (CRGDK-PEG-PCL, as active targeting model) was prepared via a maleimide-thiol reaction between CRGDK and maleimide-terminated PEG-PCL. Firstly, Mal-PEG_{41}-PCL_{17} was synthesized by ring opening polymerization of ε-caprolactone using Mal-PEG_{41}-OH as the initiator and $Sn(Oct)_2$ as the catalyst. Briefly, Mal-PEG_{41}-OH (0.4 g, 0.2 mmol), ε-caprolactone (0.387 g, 3.4 mmol) and $Sn(Oct)_2$ (0.091 mL) were placed in the dry flask. All of the reactants were dissolved in toluene (5 mL) and the reaction was placed in 100 °C oil bath. The reaction was carried out for 8 h under nitrogen atmosphere. The synthesized polymer was obtained by precipitation in ice-cooled diethyl ether. The resultant Mal-PEG_{41}-PCL_{17} was obtained after filtration and drying at 35 °C in vacuum. Subsequently, CRGDK (57.8 mg, 0.1 mmol) and Mal-PEG_{41}-PCL_{17} (0.39 g, 0.1 mmol) were directly dissolved in DMF and were then stirred for 24 h at room temperature. The product was purified by dialysis method for 24 h. CRGDK-PEG-PCL was obtained after lyophilization in a 93% yield.

The obtained DOX-*h*-PCEC and CRGDK-PEG-PCL were confirmed by using nuclear magnetic resonance (^{1}H NMR) using NMR spectrometer (VARIAN INOVA 500MHz NMR spectrometer, Varian, Palo Alto, CA, USA). DOX-*h*-PCEC and CRGDK-PEG-PCL were dissolved in dimethyl sulfoxide-d_6. The content of conjugated DOX in DOX-*h*-PCEC was quantified by UV-Vis spectrophotometer at 485 nm using WFZ-26A UV-vis spectrophotometer (Tianjin Science Instrument Plant, Tianjin, China) after hydrolyzing the hydrazone bond between DOX and PCEC under acidic conditions, according to a previous work [46]. The successful preparation of CRGDK-PEG-PCL was further confirmed by an Agilent 1100 series gel permeation chromatography (GPC) analyses (Agilent Technologies, Palo Alto, CA, USA), using Shodex GPC KF-803L column with molecular weight range 500–42,000 Da. Pure DMF was used as eluent at a flow rate of 1.0 mL/min at 30 °C.

2.3. Preparation and Characterization of CUR/DOX-h-PCEC@CRGDK NPs

The CUR/DOX codelivery nanomedicine CUR/DOX-*h*-PCEC@CRGDK NPs were prepared by solvent displacement method. Briefly, DOX-*h*-PCEC (9 mg), CRGDK-PEG-PCL (1 mg), and appropriate amount of CUR was completely dissolved in DMSO (1 mL). Then, the solution was added dropwise to 10 mL deionized water under magnetic stirring. The solution was placed in a dialysis bag (MWCO = 3.5 kDa) and dialyzed against water to remove DMSO and free substance at room temperature. With similar procedure, DOX-*h*-PCEC NPs, CUR/DOX-*h*-PCEC NPs, DOX-*h*-PCEC@CRGDK NPs, and CUR loaded CRGDK-PEG-PCL NPs were prepared by using DOX-*h*-PCEC, DOX-*h*-PCEC, together with CUR, DOX-*h*-PCEC, together with CRGDK-PEG-PCL, CRGDK-PEG-PCL, together with CUR, respectively. In addition, CUR/DOX coloaded PCEC NPs were also prepared with similar procedure by using PCEC together with different amount of CUR and DOX (PCEC:DOX:CUR in feed solution = 100:10:10, *w*/*w*/*w*).

The amount of DOX in the several NPs was determined by UV–vis spectrophotometer (Tianjin Science Instrument Plant, Tianjin, China) (absorption at 550 nm, minimizing the effect of CUR on UV measurement) using a standard curve method, according to a previous work [48]. The amount of CUR in a variety of NPs prepared as mentioned above was determined by HPLC (Agilent Technologies 1200 Series, Santa Clara, CA, USA) on a Kromasil-C18 column (4.6 mm × 250 mm, 5 μm) with a UV-vis detector at wavelength of 425 nm. 1 mg of lyophilized CUR encapsulated NPs was dissolved in 10 mL acetonitrile. It was analyzed by a mobile phase of acetonitrile and 3% acetic acid aqueous solution in the ratio of 70:30 (*v*/*v*) at a flow rate of 1 mL/min at 25 °C. Then, the amount of CUR can be achieved by the integral area of HPLC using a standard curve method. Drug loading content (DLC) and drug loading efficiency (DLE) were calculated from the following equations:

$$\text{DLC\%} = \frac{\text{weight of loaded drug}}{\text{weight of drug} - \text{loaded NPs}} \times 100\% \quad (1)$$

$$\text{DLE\%} = \frac{\text{weight of loaded drug}}{\text{weight of drug in feed}} \times 100\% \quad (2)$$

The size and size distribution as well as zeta potential of the NPs were measured by dynamic light scattering (DLS) (Malvern Zetasizer Nano ZS, Malvern, UK). The stability of NPs with different pH value was measured through sizes by DLS too. Transmission electron microscopy (TEM) was used to observe the morphology of NPs, using a JEM-100CX II instrument (JEOL LTD, Tokyo, Japan). The samples were prepared by adding few drops of the nanoparticle dispersion on the TEM copper grid. To examine the hemodynamic stability of the obtained CUR/DOX-*h*-PCEC@CRGDK NPs, we suspended CUR/DOX-*h*-PCEC@CRGDK NPs in PBS (pH 7.4) containing 5% bovine serum albumin (BSA) at 37 °C, and then the particle size change was monitored by DLS.

The encapsulation stability of CUR in CUR/DOX-*h*-PCEC@CRGDK NPs and CUR loaded PCEC NPs was comparatively studied by an ultrafiltration centrifugal tube. The CUR encapsulated NPs were diluted in PBS pH 7.4 and 5.0 containing 10% methanol as solubilizer, and then stored for 0 h, 24 h, and 48 h. After being filtered using a Millipore Amicon Ultra-0.5 centrifugal filter (MWCO $\approx$ 10 kDa, Millipore Corp., Bedford, MA, USA), the UV absorption curves of the filtrate that were collected at a different time were measured by UV-vis spectrophotometer. The changes in size of CUR/DOX-*h*-PCEC@CRGDK NPs due to acid-triggered hydrolysis of hydrazone bonds were followed by DLS. The NPs solutions in PBS pH 7.4 and 5.0 were prepared, and then the changes in NPs size at 37 °C were monitored in time by DLS.

2.4. *In Vitro Drug Release*

The release of drugs from the above NPs was studied using the dialysis method. 2 mL NPs solution was added into a dialysis bag (MWCO = 3.5 kDa) and then placed in a glass tube containing 25 mL of PBS solution. Either DOX or CUR release from NPs (dual drug loaded formulation) was investigated in PBS with different pH (pH = 6.5 and 7.4) or acetate buffer (pH = 5.0) containing 1% *v*/*v* of Tween 80. The in vitro release was placed in incubator shaker under shaking (70 rpm) at 37 °C. At the appropriate time intervals, 5 mL of the release medium was removed and replaced with equivalent fresh medium. The amount of DOX (Retention time $\approx$ 5.6 min) and CUR (Retention time $\approx$ 11.2 min) was determined by HPLC as a gradient elution method. The cumulative drug release percentage was calculated using the following equation:

$$C_r(\%) = \frac{V_t \sum_1^{n-1} C_i + V_0 C_n}{m_{drug}} \times 100\% \quad (3)$$

where C_r is the cumulative release amount; m_{drug} was the amount of drug in the formations; V_0 represented the whole volume of the release medium (V_0 = 25 mL); V_t was the volume of the replaced medium (V_t = 5.0 mL); and, C_n was the concentration of drug in the sample.

2.5. *In Vitro Cell Uptake*

The HUVEC cells and MCF-7/ADR cells were seeded into a 12-well plate at a density of 1.0×10^5 cells per well and were incubated at 37 °C for 24 h. After that, the cells were incubated with 50 nM Lyso-Tracker Green and DAPI for 30 min and 60 min, respectively (MCF-7/ADR cells were only incubated with DAPI). After washing three times with PBS, the cells were incubated with fresh medium containing of several formulations: (A) free DOX; (B) free CUR; (C) physical mixture of DOX and CUR; (D) DOX-*h*-PCEC NPs; and, (E) CUR/DOX-*h*-PCEC@CRGDK NPs. The concentration of different NPs was at an equivalent DOX dosage of 10 µg/mL. After incubation for 4 h at 37 °C, the cells were washed with PBS. The cellular uptake was visualized by confocal laser scanning microscope (Leica Microsystems, Heidelberg, Germany). The gray value was used to achieve semi-quantitative analyses by software Image-Pro Plus 6.0 (Media Cybernetics, Inc., Silver Spring, MA, USA).

2.6. *In Vitro Cytotoxicity Study*

MTT assay was performed to compare the cytotoxicity of free DOX, physical mixture of DOX and CUR, and CUR/DOX-*h*-PCEC@CRGDK NPs against MCF-7/ADR cells. MCF-7/ADR cells were seeded into the 96-well plates at 8000 cells per well, cultured in 100 µL complete dulbecco's modified eagle medium (DMEM) with the addition of DOX (1 µg/mL), and incubated at 37 °C with 5% CO_2. The supernatants were discarded, and the cells were washed twice with PBS (pH 7.4) after 24 h. Free DOX, physical mixture of DOX and CUR, and CUR/DOX-*h*-PCEC@CRGDK NPs were put in the wells for 24 h at a DOX concentration of 0.125, 1.25, 2.5, 5, 10, 20, 40 µg/mL, respectively. The culture medium without DOX (PBS) was put in the wells for control groups. Then, 20 µL MTT dye (5 mg/mL) was added. Then, the cells were incubated for 4 h at 37 °C and the medium was replaced by 150 µL

DMSO to dissolve the resulting formazan crystals. The absorption was recorded at 570 nm using a microplate reader (Thermo Scientific Varioskan Flash, Waltham, MA, USA). The experiments were conducted in triplicate and the results were presented as the average ± standard deviation.

Cell viability rate (%) was calculated as the following Equation (4):

$$\text{Cell Viability}(\%) = \frac{I_0 - I_1}{I_0} \times 100\% \quad (4)$$

I_0 was the absorbance of the cells incubated with the culture medium. I_1 was the absorbance of the cells incubated with different formulations. The median inhibitory concentration (IC_{50}) was calculated using the software GraphPad Prism 5 (GraphPad software Inc., San Diego, CA, USA).

3. Results and Discussion

3.1. Preparation and Characterization of DOX-h-PCEC and CRGDK-PEG-PCL

The lyso/endosomal pH-sensitive macromolecular prodrug NPs derived from drug molecules covalently conjugated to the polymer chains via a cleavable hydrazone can exhibit a variety of superiorities over physical drug entrapment, such as improving total drug loading, minimizing drug leakage during circulation, and intracellular rapid drug release, which have received tremendous attention for cancer chemotherapy [46,49,50]. The block copolymers composed of PEG and PCL as building block, in our opinion, should be one of the best choices for the construction of prodrug and codelivery system, due to their adjustable amphiphilicity for self-assembly into NPs, excellent biocompatibility and biodegradability, as well as stealth-shielding property of PEG [46,51,52]. Triblock copolymers PCL-PEG-PCL, i.e., PCEC, were selected in this study, offering one more terminal group for conjugation of DOX, and thus increasing total drug payload. For preparing hydrazone-linked DOX prodrug, the terminal hydroxyl groups of PCEC should be converted to hydrazide moieties, and reacted with the ketonic groups of DOX to produce DOX-*h*-PCEC. In some previous studies [46,53,54], hydrazine monohydrate was often used to activate the end groups of polymer precursor. However, we found that the partial hydrolysis of PCEC will occur when hydrazine monohydrate was used to activate PCEC. Thus, the hydrazide-functionalized PCEC was prepared by the combinational activation of NPC and tert-butyl carbazate (Boc-$NHNH_2$) and subsequent deprotection by TFA (CF_3COOH), as shown in Scheme 2. Finally, DOX-*h*-PCEC were obtained after the reaction between the hydrazide groups of PCEC and the ketonic groups of DOX. The obtained NPC-activated PCEC and hydrazide-functionalized PCEC as well as DOX-*h*-PCEC were characterized by ^{1}HNMR, as shown in Figure 1A. When compared with the ^{1}HNMR spectrum of original PCEC, no significant changes in the characteristic peaks corresponding to the PEG at about 3.6 ppm and PCL moieties at about 1.0–2.2 ppm in the ^{1}HNMR spectra of functionalized PCEC were observed. Meanwhile, the characteristic NPC peaks at about 8.1–8.3 in the spectrum of NPC-activated PCEC and the characteristic phenyl peaks of DOX at about 7.5–8.0 in the spectrum of DOX-*h*-PCEC can be observed. These results indicated that the terminal functionalized PCEC and DOX-*h*-PCEC were successfully obtained, simultaneously minimizing the PCEC hydrolysis. The peak intensities of ethylene protons (at about 3.6 ppm) of the PEG and the phenyl hydrogen protons of DOX (at about 8.0 ppm) can be used to evaluate the conjugated DOX content in DOX-*h*-PCEC. The DOX content in DOX-*h*-PCEC is about 15.2 wt % (the molar conjugation efficacy of DOX to PCEC chain is about 90%). In addition, the appearance of a strong DOX absorption peak at about 490 nm in the UV spectrum of DOX-*h*-PCEC indicated the successful conjugation of DOX to PCEC (Figure 1B). The UV peak intensity also indicated that the conjugation efficacy of DOX on PCEC was about 90%, which was well consistent with the result of ^{1}HNMR data.

CRGDK-PEG-PCL was prepared by an efficient maleimide-thiol reaction between thiol group of CRGDK and maleimide group of Mal-PEG-PCL at room temperature. To confirm the coupling between CRGDK and Mal-PEG-PCL, ^{1}HNMR was first used to analyze the product. As shown in Figure 2A, in addition of the characteristic peaks corresponding to the PEG at about 3.6 ppm and PCL moieties at about 1.0–2.2 ppm, the ^{1}HNMR spectrum of CRGDK-PEG-PCL shows the characteristic peaks corresponding to the CRGDK segment in the range of 6.5–9.0 ppm, indicating the coupling between PEG-PCL polymer chain and the CRGDK moiety. To further confirm the conjugation reaction, GPC was used to analyze the product. Both Mal-PEG-PCL and CRGDK-PEG-PCL possess a monomodal molecular weight distribution and low polydispersity. When compared to Mal-PEG-PCL, the curve of the conjugated product shifted to a lower retention time, indicating an increase in molecular weight, and thus demonstrating the successful conjugation of CRGDK to Mal-PEG-PCL.

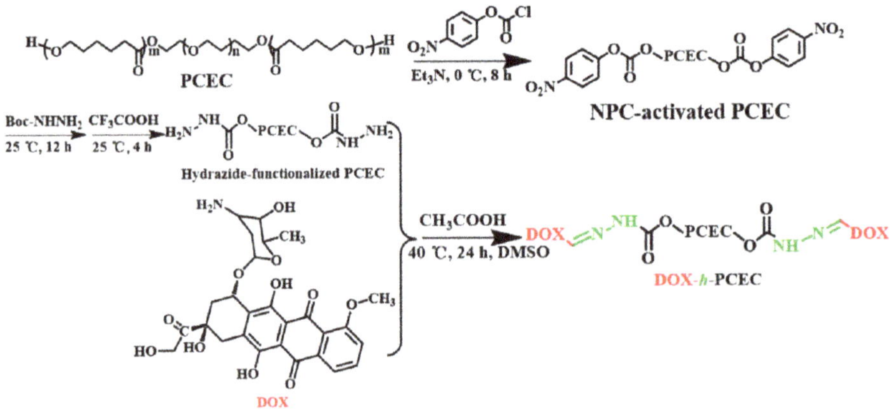

Scheme 2. Synthetic route of DOX-*h*-PCEC.

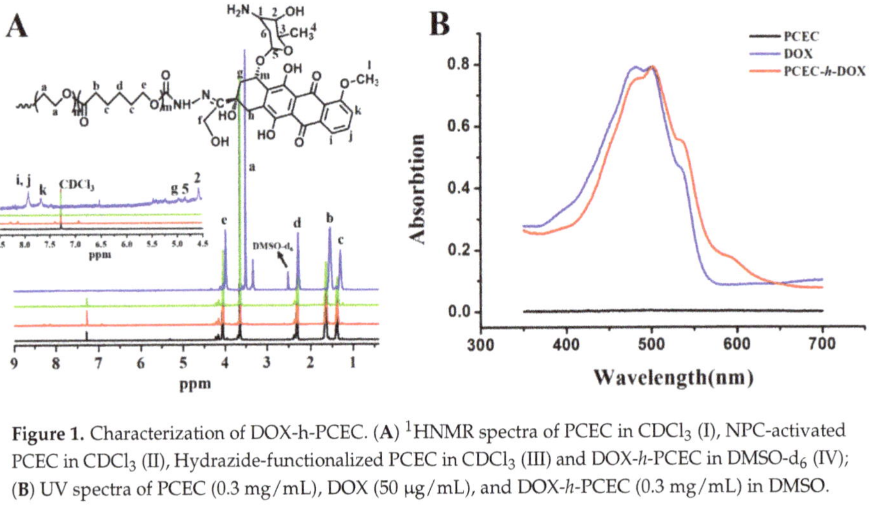

Figure 1. Characterization of DOX-h-PCEC. (**A**) ^{1}HNMR spectra of PCEC in $CDCl_3$ (I), NPC-activated PCEC in $CDCl_3$ (II), Hydrazide-functionalized PCEC in $CDCl_3$ (III) and DOX-*h*-PCEC in DMSO-d_6 (IV); (**B**) UV spectra of PCEC (0.3 mg/mL), DOX (50 μg/mL), and DOX-*h*-PCEC (0.3 mg/mL) in DMSO.

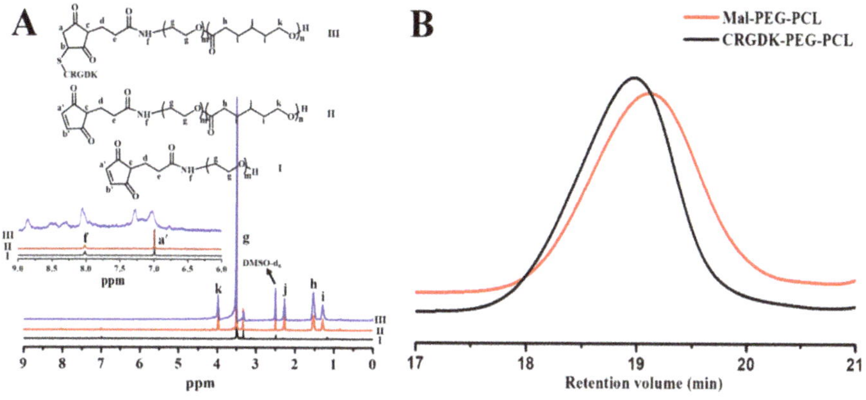

Figure 2. Characterization of CRGDK-PEG-PCL. (**A**) ^{1}HNMR spectra of Mal-PEG (I), Mal-PEG-PCL (II) and CRGDK-PEG-PCL (III) in DMSO-d_6; (**B**) gel permeation chromatography (GPC) curves of Mal-PEG-PCL and CRGDK-PEG-PCL in DMF.

3.2. Preparation and Characterization of CUR/DOX-h-PCEC@CRGDK NPs

The coassembly of two or more components has demonstrated to be a simple and efficient method to construct well-defined nanostructures with diverse structures, morphologies, and functions towards applications in drug delivery [26,55–58]. In addition, this strategy was also considered to be effective approach to ratiometrically control multidrug loading [19,26,56,57]. The amphiphilic DOX-*h*-PCEC and CRGDK-PEG-PCL cannot only self-assemble into NPs, but also coassemble into nanoformulations together with or without additional hydrophobic drugs. Apparently, the integration of DOX-*h*-PCEC as drug vector model, CRGDK-PEG-PCL as active targeting model and appropriate amount of hydrophobic CUR as MDR modulator model, via coassembly approach, will generate CUR/DOX-coloaded, CRGDK-targeted, and MDR-inhibited combinational drug delivery. A series of DOX, CUR, and CUR/DOX loaded NPs were successfully prepared by ratiometric coassembly of a different model. Their physicochemical properties were characterized in detail, as shown in Table 1. In our experimental condition, the results showed that all CUR/DOX-*h*-PCEC@CRGDK NPs showed remarkably higher encapsulation efficiencies toward CUR, especially at a relatively low CUR dose in feed solution. Even in the case of high CUR dose in feed solution (about 20%), the DLE of CUR/DOX-*h*-PCEC@CRGDK NPs is about 71.4%, which is much higher than that of CUR loaded CRGDK-PEG-PCL. These phenomena should be ascribed to the combination of hydrophobic interaction between PCL and CUR with strong intermolecular $\pi-\pi$ stacking action of DOX benzene in DOX-*h*-PCEC and CUR [34,59,60]. Moreover, as shown in the Table 1, the DLC of CUR in CUR/DOX-*h*-PCEC@CRGDK NPs can be ratiometrically controlled in a dose-depended manner. In addition, the results indicated that all of the DOX, CUR, and CUR/DOX loaded NPs by modular coassembly had low PDI of 0.11−0.23 and particle sizes ranging from 95 ± 6 to 135 ± 11 nm, depending on the NPs composition and CUR loading levels.

Table 1. Characterizations of DOX, CUR, and CUR/DOX loaded NPs by modular coassembly.

Samples	Weight Ratio in Feed Solution (DOX-*h*-PCEC:CRGDK-PEG-PCL:CUR)	Size [a] (nm)	PDI [a]	DLC of CUR (%)	DLE of CUR (%)	DOX Content [b] (wt %)	CUR Content [c] (wt %)
I [d]	90:10:0	115 ± 3	0.16	-	-	13.7	0
II	90:10:1	114 ± 5	0.11	0.93	92.7	13.7	0.92
III	90:10:2	108 ± 3	0.12	1.82	91.1	13.7	1.79
IV	90:10:5	110 ± 5	0.10	4.45	88.9	13.7	4.26
V	90:10:10	98 ± 3	0.11	8.07	80.7	13.7	7.45
VI	90:10:20	105 ± 7	0.17	14.28	71.4	13.7	12.50
VII [e]	100:0:0	118 ± 5	0.19	-	-	15.2	0
VIII [f]	100:0:10	95 ± 6	0.15	8.23	82.3	15.2	7.61
IX [g]	0:100:10	135 ± 11	0.23	3.76	37.6	0	3.62

[a] Determined by DLS; [b] Calculated by the ratio of the weight of conjugated DOX to the total weight of DOX-*h*-PCEC and CRGDK-PEG-PCL in feed; [c] Calculated by the ratio of the weight of encapsulated CUR to the weight of NPs; [d] Sample I is DOX-*h*-PCEC@CRGDK NPs; [e] Sample VII is DOX-*h*-PCEC NPs; [f] Sample VIII is CUR/DOX-*h*-PCEC NPs; [g] Sample IX is CUR loaded CRGDK-PEG-PCL NPs.

Typically, the CUR/DOX-*h*-PCEC@CRGDK NPs prepared by ratiometric coassembly of DOX-*h*-PCEC, CRGDK-PEG-PCL and CUR at a predetermined dose ratio (90:10:10, *w*/*w*/*w*), unless specifically indicated, were further characterized for their properties and performances. As shown in Figure 3A,B, CUR/DOX-*h*-PCEC@CRGDK NPs exhibited spherical morphology with a uniform size distribution. Figure 3C indicated that the CUR/DOX-*h*-PCEC@CRGDK NPs displayed a neutral zeta potential of 0.41 ± 0.12 mV, which should be conductive to avoid NPs being cleared by kidneys infiltration. The variation of particle size in BSA-containing PBS was often used to reflect the hemodynamic stability of nanocarriers. The size change as a function of incubation time for CUR/DOX-*h*-PCEC@CRGDK NPs in PBS of pH 7.4 with 5% BSA at 37 °C is shown in Figure 3D. No significant variation was observed during the entire time of observation. It indicated no occurrence of particle aggregation, and thus high colloidal stability. The good stability can be ascribed to that the steric stabilization of PEG shell. Figure 3E further indicated the physical stability of CUR/DOX-*h*-PCEC@CRGDK NPs at pH 7.4, as no any significant variation of particle size was observed over 24 h at pH 7.4. However, under otherwise the same conditions, the considerable expansion of NPs at pH 5.0 was observed. The hydrazone bonds between PCEC and DOX could be specifically hydrolyzed in the acidic environment, which would lead to the disassembly of NPs and the acceleration of the drug release. In addition, the conjugation of drug molecules to the PCL chain has been proven to be able to disturb the PCL chain regularity, hereby decreasing the crystallization of the PCL segments and enhancing the acid-triggered disassembly of NPs with PCL as the core-forming block [46,47]. The CUR/DOX-*h*-PCEC@CRGDK NPs were incubated in different pH media for a predetermined time, followed by filtration through an Amicon ultrafilter to collect the released CUR. The corresponding CUR loaded CRGDK-PEG-PCL NPs were tested in parallel as a control group. It was found that a significantly lower amount of CUR was leaked out of the CUR/DOX-*h*-PCEC@CRGDK NPs as compared to the CUR loaded CRGDK-PEG-PCL NPs at pH 7.4 (Figure 3F). Their higher capability to effectively retain encapsulated CUR indicated the greater stabilization due to the formation of stronger interaction, i.e., intermolecular $\pi-\pi$ stacking action between DOX and CUR in inner hydrophobic core of CUR/DOX-*h*-PCEC@CRGDK NPs. However, under acidic condition (pH 5.0), higher UV intensity assigned to the leakage of CUR (440 nm), and DOX (>500 nm) were observed, further demonstrating the pH-sensitiveness of CUR/DOX-*h*-PCEC@CRGDK NPs. The results also confirmed that the acid-enhanced hydrolysis of the hydrazone linkage could cause rapid and simultaneous release of DOX and CUR. Apparently, the simultaneous corelease of DOX and CUR in response to the tumor intracellular acidic microenvironment is highly desirable for the effective treatment of cancer.

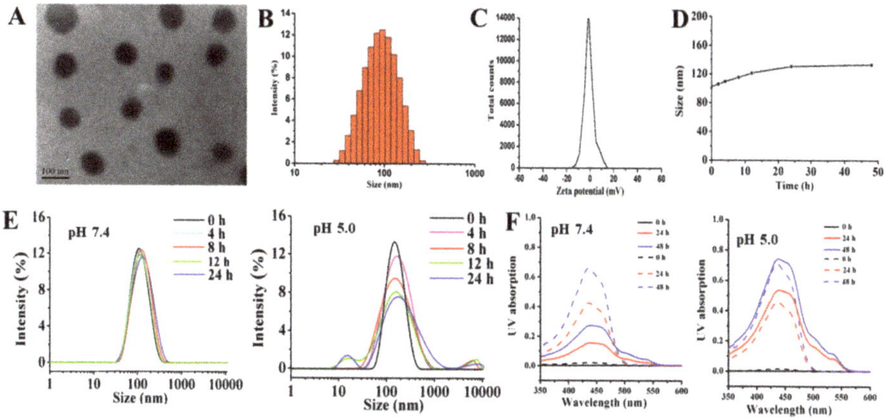

Figure 3. Characterizations of CUR/DOX-*h*-PCEC@CRGDK NPs. (**A**) TEM image; (**B**) size histogram; (**C**) Zeta potential at pH 7.4; (**D**) Hemodynamic stability in pH 7.4 PBS containing 5% BSA at 37 °C; (**E**) Size changes of NPs incubated in PBS of pH 7.4 and pH 5.0; (**F**) CUR leakage from CUR/DOX-*h*-PCEC@CRGDK NPs (solid line) and CUR loaded CRGDK-PEG-PCL NPs (dashed line) incubated in PBS of pH 7.4 and pH 5.0 containing 10% methanol as solubilizer; The leaked CUR samples were diluted 10 times by same incubation solution.

3.3. In Vitro Drug Release

The in vitro release profiles of DOX and CUR from CUR/DOX-*h*-PCEC@CRGDK NPs and CUR/DOX coloaded PCEC NPs as control were comparatively studied at 37 °C at pH 7.4 (the extracellular pH of normal tissue), 6.5 (the extracellular environment of tumor tissues), and 5.0 (lyso/endosomal environment of tumor cells) using a dialysis method, as shown in Figure 4. Both the release rate of DOX and CUR from CUR/DOX coloaded PCEC NPs at pH 7.4 was relatively low, parallel to each other throughout the study (Figure 4A,C). Only approximately 15% of drug was released from PCEC NPs within 24 h. With the decrease of the pH values, no marked change for CUR release from PCEC NPs (Figure 4C). However, when compared with the CUR release from PCEC NPs, the release of DOX from PCEC NPs was slightly accelerated at pH 6.5 and pH 5.0, which may be due to the improved solubility of DOX because of DOX protonation in acidic media. The effects of pH values on the in vitro release behaviors of DOX from CUR/DOX-*h*-PCEC@CRGDK NPs are shown in Figure 4B. A significantly pH-dependent release profile was observed. The release of conjugated DOX from CUR/DOX-*h*-PCEC@CRGDK NPs was very low, at pH 7.4. Only approximately 6% of DOX was released at 24 h, which was much lower than DOX release from PCEC NPs (about 15%). These results can be due to the high stability of hydrazone bonds between DOX and polymers under physiological conditions (pH 7.4). However, the release rate of DOX was significantly increased at acidic pH values. The DOX was released about 58.6% at pH 6.5 and 78.2% at pH 5.0 within 24 h, respectively. The significant increase in DOX release can be attributed to the combination of acid-induced hydrolysis of hydrazone bond, accelerated disassembly of NPs, and enhanced solubility of DOX in acidic media. Similarly, a pH-controlled release profile of CUR from CUR/DOX-*h*-PCEC@CRGDK NPs was observed (Figure 4D). The CUR release rate was significantly faster at pH 6.5 and 5.0. About 70% of the CUR was released from CUR/DOX-*h*-PCEC@CRGDK NPs within 24 h. The significant increase in CUR release under acidic conditions can be ascribed to that the pH-induced hydrolysis of hydrazone bond can cause the swelling and disassembly of CUR/DOX-*h*-PCEC@CRGDK NPs, as demonstrated in Figure 3E,F, facilitating the diffusion and release of CUR. These results further indicated that the simultaneous corelease of DOX and CUR by acid-triggered degradation of prodrug and subsequent disassembly of NPs occurred, which will be desirable to maximize the synergistic effect of DOX and

CUR within the tumor cells. In sum, CUR/DOX-*h*-PCEC@CRGDK NPs presented pH-dependent DOX and CUR corelease behaviors, which should be beneficial for tumor treatment. Most of DOX and CUR will remain in NPs for a considerable length of time at normal physiological conditions (pH 7.4). However, both the loaded DOX and CUR will be rapidly and simultaneously released due to the acidic environment in lyso/endocytic compartments (pH 5.0–6.5) after uptake by tumor cells.

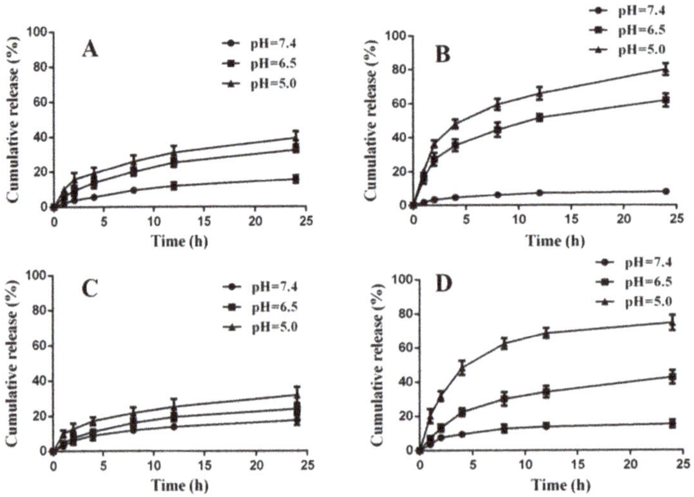

Figure 4. In vitro drug release profiles. (**A**) DOX release from CUR/DOX coloaded PCEC NPs; (**B**) DOX release from CUR/DOX-*h*-PCEC@CRGDK NPs; (**C**) CUR release from CUR/DOX coloaded PCEC NPs; (**D**) CUR release from CUR/DOX-*h*-PCEC@CRGDK NPs;.

3.4. In Vitro Cell Uptake

Efficient and selective cellular uptake of therapeutic drugs plays a key role in improving chemotherapy efficacy and decreasing side effects. Neuropilin-1 is overexpressed in many human cancer cells, providing a useful targeting site for tumor-specific drug delivery [41–45]. In this study, the cellular uptake of free DOX, DOX-*h*-PCEC NPs, and CUR/DOX-*h*-PCEC@CRGDK NPs was monitored by confocal laser scanning microscope in Neuropilin-1 overexpressed HUVEC cells. DAPI was used to label nucleus (blue) and Lyso-Tracker was applied to label lysosomes (green). As shown in Figure 5A, the red signals, due to the intrinsic fluorescence of DOX, were clearly observed in the cytoplasm and nuclei of the treated HUVEC cells, indicating the rapid passive diffusion of free DOX and effective endocytosis of DOX-loaded NPs. Further inspection of Figure 5A shows that, for the cells after 4 h of incubation with free DOX, the relative weak fluorescence was observed in the cytoplasm of cells, indicating that the DOX molecules successfully entered the cells, but mainly accumulated in the cytoplasm. By contrast, when the cells were incubated with DOX-loaded NPs, the fluorescence mainly appeared in the proximity of cellular nuclei. When compared with fluorescence intensity in the cells that were treated with DOX-*h*-PCEC NPs, the detected fluorescence in the cells treated with CUR/DOX-*h*-PCEC@CRGDK NPs was much higher in the nuclei. Apparently, significantly enhanced cellular uptake and considerably increased nucleus localization were confirmed in HUVEC cells that were incubated with CUR/DOX-*h*-PCEC@CRGDK NPs. Moreover, as a measurement of fluorescence intensity of DOX, the gray value for images, as a gross approximation, was determined using the image analysis software (Figure 5B). The results of measuring fluorescence intensity based on average gray value also demonstrated that both the cell uptake and nucleus localization of CUR/DOX-*h*-PCEC@CRGDK NPs were much higher than that of free DOX and DOX-*h*-PCEC NPs without CRGDK. These results definitely revealed that CRGDK surface decorated CUR/DOX-*h*-PCEC@CRGDK NPs possess significantly improved binding activity with Neuropilin-1 overexpressing cancer cells, and thus can exploit the efficient receptor-mediated endocytosis to achieve cell-specific drug delivery.

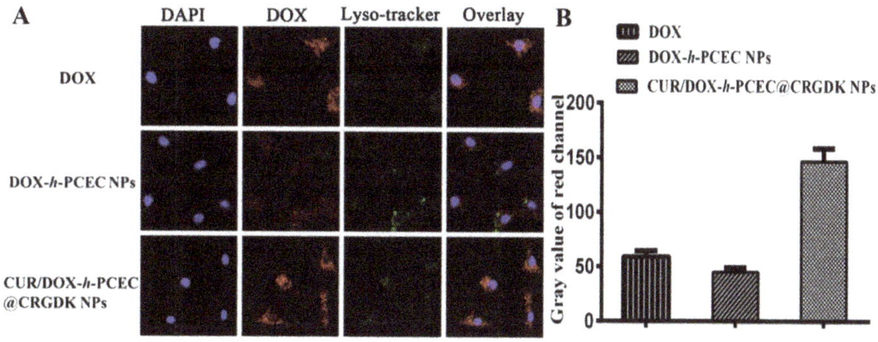

Figure 5. Selective cellular uptake. (**A**) Representative fluorescence microscopy images of Neuropilin-1 overexpressed human umbilical vein endothelial cells (HUVEC) cells incubated with free DOX, DOX-*h*-PCEC NPs, and CUR/DOX-*h*-PCEC@CRGDK NPs at equivalent DOX concentration of 10 μg/mL for 4 h; (**B**) Average gray value as a criterion to compare the red fluorescent intensity in different images.

3.5. Intracellular Synergistic Corelease and Cytotoxicity

As is well known, the MDR cells possess drug efflux abilities by overexpressing P-gp, presenting a critical challenge to effective cancer therapy [5,8]. In the present study, to verify whether CUR/DOX-*h*-PCEC@CRGDK NPs enabled cells to overcome MDR by synergistic corelease of DOX and CUR to inhibit P-gp overexpression, we applied the intrinsic fluorescence of DOX and CUR to carry out a comparative analysis of intracellular drug localization in the drug-resistant MCF-7/ADR cells treated with free DOX, free CUR, DOX+CUR (1.0:1.0, *w*/*w*), and CUR/DOX-*h*-PCEC@CRGDK NPs. As shown in Figure 6A, no obvious intracellular accumulation of DOX within the MCF-7/ADR cells treated with free DOX was observed, presumably due to a powerful MDR effect. It can be seen that the DOX + CUR group exhibited higher intracellular DOX accumulation than free DOX group, confirming that CUR can facilitate DOX uptake into MCF-7/ADR cells. More importantly, the results in Figure 6A clearly showed that, in comparison with free DOX and DOX+CUR group, significantly stronger DOX and CUR fluorescence appeared within the cells treated by CUR/DOX-*h*-PCEC@CRGDK NPs. Notably, CUR/DOX-*h*-PCEC@CRGDK NPs can significantly increase the cellular uptake of both DOX and CUR in MCF-7/ADR cells, which can be attributed to the effective internalization because of the active targeting effect of CRGDK and efficient intracellular corelease due to the acid-triggered cleavage of hydrazone. In addition, these results not only suggested that the CUR/DOX-*h*-PCEC@CRGDK NPs can synergistically and simultaneously corelease DOX and CUR, but also confirmed that the nanoparticulate codelivery of DOX and CUR can more effectively inhibit the overexpression of P-gp and thus reverse the MDR, finally increasing the intracellular accumulation and retention of DOX.

To demonstrate the in vitro anticancer activity of CUR/DOX-*h*-PCEC@CRGDK NPs, we tested the cell cytotoxicity of free DOX, DOX+CUR (1.0:1.0, *w*/*w*), and CUR/DOX-*h*-PCEC@CRGDK NPs against MCF-7/ADR cells. All of the formulations gradually increased their cell cytotoxicity with the increase of concentrations. However, MCF-7/ADR cells showed the lowest sensitivity to the treatment with free DOX. The half-maximal inhibitory concentration (IC_{50}) value of free DOX was measured to be about 34.7 μg/mL. The relatively high IC_{50} should be due to that MCF-7/ADR cells were highly resistant to DOX. However, DOX+CUR group had a much higher cytotoxicity to MCF-7/ADR cells (IC_{50} = 12.1 ± 0.7 μg/mL) than free DOX, due to the chmosensitization effect on

MDR reversal of CUR, as mentioned above. In particular, the CUR/DOX-*h*-PCEC@CRGDK NPs presented remarkably high cytotoxicity against MCF-7/ADR cells, as their IC_{50} value was considerably decreased to 4.1 ± 0.2 μg/mL.

The CUR/DOX-*h*-PCEC@CRGDK NPs effectively corelease the encapsulated chemotherapeutic drug and chemosensitizer, thus achieving an enhanced cytotoxicity against MCF-7/ADR cells. These results further confirmed the excellent capability of CUR/DOX-*h*-PCEC@CRGDK NPs for efficient tumor targeting delivery and intracellular synergistic corelease to modulate the drug resistance.

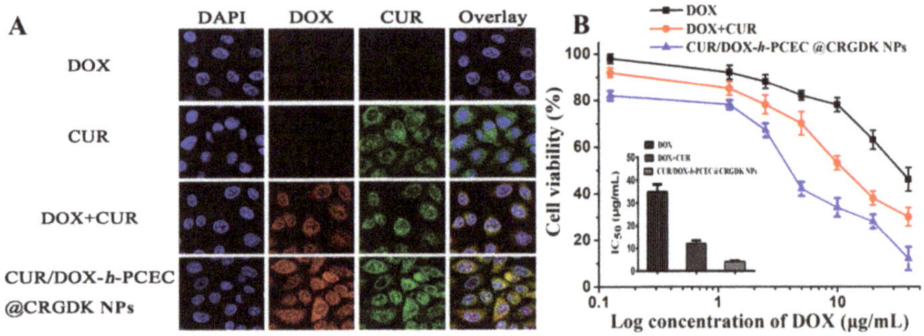

Figure 6. Intracellular synergistic corelease and cytotoxicity. (**A**) Representative fluorescence microscopy images of Adriamycin resistant MCF-7/ADR cells with free DOX, free CUR, physical mixture of DOX and CUR (1.0:1.0, *w*/*w*), and CUR/DOX-*h*-PCEC@CRGDK NPs at equivalent DOX concentration of 10 μg/mL for 4 h; (**B**) Cytotoxicity of free DOX, physical mixture of DOX and CUR (1.0:1.0, *w*/*w*), and CUR/DOX-*h*-PCEC@CRGDK NPs to MCF-7/ADR cells.

4. Conclusions

In this work, we applied a modular coassembly of acid-cleavable PEGylated polymeric prodrug, tumor cell-specific targeting peptide, and natural chemosensitizer to construct an all-in-one multifunctional multidrug delivery system (CUR/DOX-*h*-PCEC@CRGDK NPs) for the synergistic codelivery of DOX and CUR. The CUR/DOX-*h*-PCEC@CRGDK NPs were demonstrated to be able to ratiometrically load DOX and CUR, allow for long circulation, enter the cells via CRGDK-receptor mediated tumor targeting, as well as realize tumor intracellular responsive and simultaneous corelease of DOX and CUR, thus effectively reversing the MDR through inhibiting P-gp expression. We believe that the design presented here can provide a facile and robust nanoplatform for targeted multidrug codelivery to maximize synergistic effects, thus exhibiting great potential for future applications in the combinatory cancer therapy.

Acknowledgments: This work was supported by National Natural Science Foundation of China (31470925, and 31470963) and Tianjin Research Program of Application Foundation and Advanced Technology (15JCQNJC03000). National Basic Research Program of China (No. 2014CB643305) and the Research Fund of State Key Laboratory for Marine Corrosion and Protection of Luoyang Ship Material Research Institute under Contract No. KF160401.

Author Contributions: J.Z. and A.D. conceived and designed the tasks and experiments; M.Y. and L.Y. performed the experiments; R.G. and C.L. discussed the project and analyzed data; J.Z. and M.Y. wrote the manuscript. All authors reviewed and approved the manuscript.

Conflicts of Interest: The authors declare no conflict of interest.

References

1. Siegel, R.L.; Miller, K.D.; Jemal, A. Cancer statistics, 2017. *CA Cancer J. Clin.* **2017**, *67*, 7–30. [CrossRef] [PubMed]
2. Iacovelli, R.; Pietrantonio, F.; Maggi, C.; BraudaMari, F.; Bartolomeo, D. Combination or single-agent chemotherapy as adjuvant treatment of gastric cancer: A systematic review and meta-analysis of published trials. *Crit. Rev. Oncol. Hematol.* **2016**, *98*, 24–28. [CrossRef] [PubMed]
3. Shi, J.; Kantoff, P.W.; Wooster, R.; Farokhzad, O.C. Cancer nanomedicine: Progress, challenges and opportunities. *Nat. Rev. Cancer* **2017**, *17*, 20–37. [CrossRef] [PubMed]
4. Blanco, E.; Shen, H.; Ferrari, M. Principles of nanoparticle design for overcoming biological barriers to drug delivery. *Nat. Biotechnol.* **2015**, *33*, 941–951. [CrossRef] [PubMed]
5. Kunjachan, S.; Rychlik, B.; Storm, G.; Kiessling, F.; Lammers, T. Multidrug resistance: Physiological principles and nanomedical solutions. *Adv. Drug Deliv. Rev.* **2013**, *65*, 1852–1865. [CrossRef] [PubMed]
6. Jia, J.; Zhu, F.; Ma, X.; Cao, Z.; Cao, Z.W.; Li, Y.; Li, Y.X.; Chen, Y.Z. Mechanisms of drug combinations: Interaction and network perspectives. *Nat. Rev. Drug Discov.* **2009**, *8*, 111–128. [CrossRef] [PubMed]
7. Kolishetti, N.; Dhar, S.; Valencia, P.M.; Lin, L.Q.; Karnik, R.; Lippard, S.J.; Langer, R.; Farokhzad, O.C. Engineering of self-assembled nanoparticle platform for precisely controlled combination drug therapy. *Proc. Natl. Acad. Sci. USA* **2010**, *107*, 17939–17944. [CrossRef] [PubMed]
8. Chen, Z.; Shi, T.; Zhang, L.; Zhu, P.; Deng, M.; Huang, C.; Hu, T.; Jiang, L.; Li, J. Mammalian drug efflux transporters of the atp binding cassette (abc) family in multidrug resistance: A review of the past decade. *Cancer Lett.* **2016**, *370*, 153–164. [CrossRef] [PubMed]
9. Hu, Q.; Sun, W.; Wang, C.; Gu, Z. Recent advances of cocktail chemotherapy by combination drug delivery systems. *Adv. Drug Deliv. Rev.* **2016**, *98*, 19–34. [CrossRef] [PubMed]
10. Jang, B.; Kwon, H.; Katila, P.; Lee, S.J.; Lee, H. Dual delivery of biological therapeutics for multimodal and synergistic cancer therapies. *Adv. Drug Deliv. Rev.* **2016**, *98*, 113–133. [CrossRef] [PubMed]
11. Lane, D. Designer combination therapy for cancer. *Nat. Biotechnol.* **2006**, *24*, 163–164. [CrossRef] [PubMed]
12. Yan, G.; Li, A.; Zhang, A.; Sun, Y.; Liu, J. Polymer-based nanocarriers for co-delivery and combination of diverse therapies against cancers. *Nanomaterials* **2018**, *8*. [CrossRef] [PubMed]
13. Oh, H.R.; Jo, H.Y.; Park, J.S.; Kim, D.E.; Cho, J.Y.; Kim, P.H.; Kim, K.S. Galactosylated liposomes for targeted co-delivery of doxorubicin/vimentin sirna to hepatocellular carcinoma. *Nanomaterials* **2016**, *6*. [CrossRef] [PubMed]
14. Hu, C.M.; Aryal, S.; Zhang, L. Nanoparticle-assisted combination therapies for effective cancer treatment. *Ther. Deliv* **2010**, *1*, 323–334. [CrossRef] [PubMed]
15. Ma, L.; Kohli, M.; Smith, A. Nanoparticles for combination drug therapy. *ACS Nano* **2013**, *7*, 9518–9525. [CrossRef] [PubMed]
16. Kemp, J.A.; Shim, M.S.; Heo, C.Y.; Kwon, Y.J. "Combo" Nanomedicine: Co-delivery of multi-modal therapeutics for efficient, targeted, and safe cancer therapy. *Adv. Drug Deliv. Rev.* **2016**, *98*, 3–18. [CrossRef] [PubMed]
17. Jhaveri, A.; Deshpande, P.; Torchilin, V. Stimuli-sensitive nanopreparations for combination cancer therapy. *J. Control. Release* **2014**, *190*, 352–370. [CrossRef] [PubMed]
18. Hashemi, M.; Ebrahimian, M. Recent advances in nanoformulations for co-delivery of curcumin and chemotherapeutic drugs. *Nanomed. J.* **2017**, *4*, 1–7.
19. Chen, C.; Tao, R.; Ding, D.; Kong, D.; Fan, A.; Wang, Z.; Zhao, Y. Ratiometric co-delivery of multiple chemodrugs in a single nanocarrier. *Eur. J. Pharm. Sci.* **2017**, *107*, 16–23. [CrossRef] [PubMed]
20. Klippstein, R.; Bansal, S.S.; Al-Jamal, K.T. Doxorubicin enhances curcumin's cytotoxicity in human prostate cancer cells in vitro by enhancing its cellular uptake. *Int. J. Pharm.* **2016**, *514*, 169–175. [CrossRef] [PubMed]
21. Barui, S.; Saha, S.; Mondal, G.; Haseena, S.; Chaudhuri, A. Simultaneous delivery of doxorubicin and curcumin encapsulated in liposomes of pegylated rgdk-lipopeptide to tumor vasculature. *Biomaterials* **2014**, *35*, 1643–1656. [CrossRef] [PubMed]
22. Tefas, L.R.; Sylvester, B.; Tomuta, I.; Sesarman, A.; Licarete, E.; Banciu, M.; Porfire, A. Development of antiproliferative long-circulating liposomes co-encapsulating doxorubicin and curcumin, through the use of a quality-by-design approach. *Drug Des. Dev. Ther.* **2017**, *11*, 1605–1621. [CrossRef] [PubMed]

23. Zhao, X.; Chen, Q.; Liu, W.; Li, Y.; Tang, H.; Liu, X.; Yang, X. Codelivery of doxorubicin and curcumin with lipid nanoparticles results in improved efficacy of chemotherapy in liver cancer. *Int. J. Nanomed.* **2015**, *10*, 257–270.
24. Zhao, X.; Chen, Q.; Li, Y.; Tang, H.; Liu, W.; Yang, X. Doxorubicin and curcumin co-delivery by lipid nanoparticles for enhanced treatment of diethylnitrosamine-induced hepatocellular carcinoma in mice. *Eur. J. Pharm. Biopharm.* **2015**, *93*, 27–36. [CrossRef] [PubMed]
25. Sarisozen, C.; Dhokai, S.; Tsikudo, E.G.; Luther, E.; Rachman, I.M.; Torchilin, V.P. Nanomedicine based curcumin and doxorubicin combination treatment of glioblastoma with scfv-targeted micelles: In vitro evaluation on 2D and 3D tumor models. *Eur. J. Pharm. Biopharm.* **2016**, *108*, 54–67. [CrossRef] [PubMed]
26. Ma, W.; Guo, Q.; Li, Y.; Wang, X.; Wang, J.; Tu, P. Co-assembly of doxorubicin and curcumin targeted micelles for synergistic delivery and improving anti-tumor efficacy. *Eur. J. Pharm. Biopharm.* **2017**, *112*, 209–223. [CrossRef] [PubMed]
27. Wang, J.; Ma, W.; Tu, P. Synergistically improved anti-tumor efficacy by co-delivery doxorubicin and curcumin polymeric micelles. *Macromol. Biosci.* **2015**, *15*, 1252–1261. [CrossRef] [PubMed]
28. Zhang, Y.; Yang, C.; Wang, W.; Liu, J.; Liu, Q.; Huang, F.; Chu, L.; Gao, H.; Li, C.; Kong, D.; et al. Co-delivery of doxorubicin and curcumin by ph-sensitive prodrug nanoparticle for combination therapy of cancer. *Sci. Rep.* **2016**, *6*, 21225. [CrossRef] [PubMed]
29. Gao, C.; Tang, F.; Gong, G.; Zhang, J.; Hoi, M.P.M.; Lee, S.M.Y.; Wang, R. Ph-responsive prodrug nanoparticles based on a sodium alginate derivative for selective co-release of doxorubicin and curcumin into tumor cells. *Nanoscale* **2017**, *9*, 12533–12542. [CrossRef] [PubMed]
30. Cui, T.; Zhang, S.; Sun, H. Co-delivery of doxorubicin and ph-sensitive curcumin prodrug by transferrin-targeted nanoparticles for breast cancer treatment. *Oncol. Rep.* **2017**, *37*, 1253–1260. [CrossRef] [PubMed]
31. Li, W.M.; Chiang, C.S.; Huang, W.C.; Su, C.W.; Chiang, M.Y.; Chen, J.Y.; Chen, S.Y. Amifostine-conjugated ph-sensitive calcium phosphate-covered magnetic-amphiphilic gelatin nanoparticles for controlled intracellular dual drug release for dual-targeting in her-2-overexpressing breast cancer. *J. Control. Release* **2015**, *220*, 107–118. [CrossRef] [PubMed]
32. Fang, J.H.; Lai, Y.H.; Chiu, T.L.; Chen, Y.Y.; Hu, S.H.; Chen, S.Y. Magnetic core-shell nanocapsules with dual-targeting capabilities and co-delivery of multiple drugs to treat brain gliomas. *Adv. Healthc. Mater.* **2014**, *3*, 1250–1260. [CrossRef] [PubMed]
33. Dutta, B.; Shetake, N.G.; Barick, B.K.; Barick, K.C.; Pandey, B.N.; Priyadarsini, K.I.; Hassan, P.A. Ph sensitive surfactant-stabilized fe3o4 magnetic nanocarriers for dual drug delivery. *Colloids Surf. B Biointerfaces* **2018**, *162*, 163–171. [CrossRef] [PubMed]
34. Zhang, P.; Li, J.; Ghazwani, M.; Zhao, W.; Huang, Y.; Zhang, X.; Venkataramanan, R.; Li, S. Effective co-delivery of doxorubicin and dasatinib using a peg-fmoc nanocarrier for combination cancer chemotherapy. *Biomaterials* **2015**, *67*, 104–114. [CrossRef] [PubMed]
35. Wu, L.; Zhang, J.; Watanabe, W. Physical and chemical stability of drug nanoparticles. *Adv. Drug Deliv. Rev.* **2011**, *63*, 456–469. [CrossRef] [PubMed]
36. Bertrand, N.; Wu, J.; Xu, X.; Kamaly, N.; Farokhzad, O.C. Cancer nanotechnology: The impact of passive and active targeting in the era of modern cancer biology. *Adv. Drug Deliv. Rev.* **2014**, *66*, 2–25. [CrossRef] [PubMed]
37. Fang, J.; Nakamura, H.; Maeda, H. The EPR effect: Unique features of tumor blood vessels for drug delivery, factors involved, and limitations and augmentation of the effect. *Adv. Drug Deliv. Rev.* **2011**, *63*, 136–151. [CrossRef] [PubMed]
38. Sugahara, K.N.; Teesalu, T.; Karmali, P.P.; Kotamraju, V.R.; Agemy, L.; Greenwald, D.R.; Ruoslahti, E. Coadministration of a tumor-penetrating peptide enhances the efficacy of cancer drugs. *Science* **2010**, *328*, 1031–1035. [CrossRef] [PubMed]
39. Zhang, Y.; Huang, F.; Ren, C.; Yang, L.; Liu, J.; Cheng, Z.; Chu, L.; Liu, J. Targeted chemo-photodynamic combination platform based on the dox prodrug nanoparticles for enhanced cancer therapy. *ACS Appl. Mater. Interfaces* **2017**, *9*, 13016–13028. [CrossRef] [PubMed]
40. Yang, C.L.; Chen, J.P.; Wei, K.C.; Chen, J.Y.; Huang, C.W.; Liao, Z.X. Release of doxorubicin by a folate-grafted, chitosan-coated magnetic nanoparticle. *Nanomaterials* **2017**, *7*. [CrossRef] [PubMed]

41. Zhou, G.; Xu, Y.; Chen, M.; Cheng, D.; Shuai, X. Tumor-penetrating peptide modified and ph-sensitive polyplexes for tumor targeted sirna delivery. *Polym. Chem.* **2016**, *7*, 3857–3863. [CrossRef]
42. Wei, T.; Liu, J.; Ma, H.; Cheng, Q.; Huang, Y.; Zhao, J.; Huo, S.; Xue, X.; Liang, Z.; Liang, X.J. Functionalized nanoscale micelles improve drug delivery for cancer therapy in vitro and in vivo. *Nano Lett.* **2013**, *13*, 2528–2534. [CrossRef] [PubMed]
43. Kumar, A.; Huo, S.; Zhang, X.; Liu, J.; Tan, A.; Li, S.; Jin, S.; Xue, X.; Zhao, Y.; Ji, T.; et al. Neuropilin-1-targeted gold nanoparticles enhance therapeutic efficacy of platinum(iv) drug for prostate cancer treatment. *ACS Nano* **2014**, *8*, 4205–4220. [CrossRef] [PubMed]
44. Fan, X.; Zhang, W.; Hu, Z.; Li, Z. Facile synthesis of rgd-conjugated unimolecular micelles based on a polyester dendrimer for targeting drug delivery. *J. Mater. Chem. B* **2017**, *5*, 1062–1072. [CrossRef]
45. Kunjachan, S.; Pola, R.; Gremse, F.; Theek, B.; Ehling, J.; Moeckel, D.; Hermanns-Sachweh, B.; Pechar, M.; Ulbrich, K.; Hennink, W.E.; et al. Passive versus active tumor targeting using RGD- and NGR-modified polymeric nanomedicines. *Nano Lett.* **2014**, *14*, 972–981. [CrossRef] [PubMed]
46. Zhang, J.; Lin, X.; Liu, J.; Zhao, J.; Dong, H.; Deng, L.; Liu, J.; Dong, A. Sequential thermo-induced self-gelation and acid-triggered self-release process of drug-conjugated nanoparticles: A strategy for the sustained and controlled drug delivery to tumors. *J. Mater. Chem. B* **2013**, *1*, 4667–4677. [CrossRef]
47. Lin, X.; Deng, L.; Xu, Y.; Dong, A. Thermosensitive in situ hydrogel of paclitaxel conjugated poly(ε-caprolactone)-poly(ethylene glycol)-poly(ε-caprolactone). *Soft Matter* **2012**, *8*, 3470–3477. [CrossRef]
48. Maity, A.R.; Chakraborty, A.; Mondal, A.; Jana, N.R. Carbohydrate coated, folate functionalized colloidal graphene as a nanocarrier for both hydrophobic and hydrophilic drugs. *Nanoscale* **2014**, *6*, 2752–2758. [CrossRef] [PubMed]
49. Duncan, R. Polymer conjugates as anticancer nanomedicines. *Nat. Rev. Cancer* **2006**, *6*, 688–701. [CrossRef] [PubMed]
50. Deng, L.; Dong, H.; Dong, A.; Zhang, J. A strategy for oral chemotherapy via dual ph-sensitive polyelectrolyte complex nanoparticles to achieve gastric survivability, intestinal permeability, hemodynamic stability and intracellular activity. *Eur. J. Pharm. Biopharm.* **2015**, *97*, 107–117. [CrossRef] [PubMed]
51. Deng, H.; Liu, J.; Zhao, X.; Zhang, Y.; Liu, J.; Xu, S.; Deng, L.; Dong, A.; Zhang, J. PEG-b-PCL copolymer micelles with the ability of pH-controlled negative-to-positive charge reversal for intracellular delivery of doxorubicin. *Biomacromolecules* **2014**, *15*, 4281–4292. [CrossRef] [PubMed]
52. Rainbolt, E.A.; Washington, K.E.; Biewer, M.C.; Stefan, M.C. Recent developments in micellar drug carriers featuring substituted poly(ε-caprolactone)s. *Polym. Chem.* **2015**, *6*, 2369–2381. [CrossRef]
53. Hu, X.; Liu, S.; Huang, Y.; Chen, X.; Jing, X. Biodegradable block copolymer-doxorubicin conjugates via different linkages: Preparation, characterization, and in vitro evaluation. *Biomacromolecules* **2010**, *11*, 2094–2102. [CrossRef] [PubMed]
54. Rosario, L.S.D.; Demirdirek, B.; Harmon, A.; Orban, D.; Uhrich, K.E. Micellar nanocarriers assembled from doxorubicin-conjugated amphiphilic macromolecules (dox–am). *Macromol. Biosci.* **2010**, *10*, 415–423. [CrossRef] [PubMed]
55. Liu, C.; Li, M.; Yang, J.; Xiong, L.; Sun, Q. Fabrication and characterization of biocompatible hybrid nanoparticles from spontaneous co-assembly of casein/gliadin and proanthocyanidin. *Food Hydrocoll.* **2017**, *73*, 74–89. [CrossRef]
56. Hu, C.M.; Zhang, L. Nanoparticle-based combination therapy toward overcoming drug resistance in cancer. *Biochem. Pharmacol.* **2012**, *83*, 1104–1111. [CrossRef] [PubMed]
57. Aryal, S.; Hu, C.-M.J.; Zhang, L. Polymeric nanoparticles with precise ratiometric control over drug loading for combination therapy. *Mol. Pharm.* **2011**, *8*, 1401–1407. [CrossRef] [PubMed]
58. Yang, Y.; Zhang, Y.-M.; Li, D.; Sun, H.-L.; Fan, H.-X.; Liu, Y. Camptothecin–polysaccharide co-assembly and its controlled release. *Bioconjug. Chem.* **2016**, *27*, 2834–2838. [CrossRef] [PubMed]

59. Zhao, J.; Wang, H.; Liu, J.; Deng, L.; Liu, J.; Dong, A.; Zhang, J. Comb-like amphiphilic copolymers bearing acetal-functionalized backbones with the ability of acid-triggered hydrophobic-to-hydrophilic transition as effective nanocarriers for intracellular release of curcumin. *Biomacromolecules* **2013**, *14*, 3973–3984. [CrossRef] [PubMed]
60. Yan, T.; Li, D.; Li, J.; Cheng, F.; Cheng, J.; Huang, Y.; He, J. Effective co-delivery of doxorubicin and curcumin using a glycyrrhetinic acid-modified chitosan-cystamine-poly(epsilon-caprolactone) copolymer micelle for combination cancer chemotherapy. *Colloids Surf. B Biointerfaces* **2016**, *145*, 526–538. [CrossRef] [PubMed]

nanomaterials

MDPI

Article

Spherical and Spindle-Like Abamectin-Loaded Nanoparticles by Flash Nanoprecipitation for Southern Root-Knot Nematode Control: Preparation and Characterization

Zhinan Fu [1], Kai Chen [1,2], Li Li [1], Fang Zhao [1], Yan Wang [3], Mingwei Wang [1], Yue Shen [3,*], Haixin Cui [3,*], Dianhua Liu [1] and Xuhong Guo [1,2,*]

1 State Key Laboratory of Chemical Engineering, East China University of Science and Technology, Shanghai 200237, China; fzn940223@163.com (Z.F.); chenkai@shzu.edu.cn (K.C.); lili76131@ecust.edu.cn (L.L.); fzhao1@ecust.edu.cn (F.Z.); mingweiwang@ecust.edu.cn (M.W.); dhliu@ecust.edu.cn (D.L.)

2 Engineering Research Center of Materials Chemical Engineering of Xinjiang Bingtuan, Shihezi University, Xinjiang 832000, China

3 Environment and Sustainable Development in Agriculture, Chinese Academy of Agricultural Sciences, Beijing 100081, China; wangyan03@caas.cn

* Correspondence: shenyue@caas.cn (Y.S.); cuihaixin@caas.cn (H.C.); guoxuhong@ecust.edu.cn (X.G.); Tel.: +86-10-8210-5997 (Y.S.); +86-10-8210-6013 (H.C.); +86-21-6425-3491 (X.G.)

Received: 30 May 2018; Accepted: 18 June 2018; Published: 20 June 2018

Abstract: Southern root-knot nematode (*Meloidogyne incognita*) is a biotrophic parasite, causing enormous loss in global crop production annually. Abamectin (Abm) is a biological and high-efficiency pesticide against *Meloidogyne incognita*. In this study, a powerful method, flash nanoprecipitation (FNP), was adopted to successfully produce Abm-loaded nanoparticle suspensions with high drug loading capacity (>40%) and encapsulation efficiency (>95%), where amphiphilic block copolymers (BCPs) poly(lactic-co-glycolic acid)-*b*-poly(ethylene glycol) (PLGA-*b*-PEG), poly(D,L-lactide)-*b*-poly(ethylene glycol) (PLA-*b*-PEG), or poly(caprolactone)-*b*-poly(ethylene glycol) (PCL-*b*-PEG) were used as the stabilizer to prevent the nanoparticles from aggregation. The effect of the drug-to-stabilizer feed ratio on the particle stability were investigated. Moreover, the effect of the BCP composition on the morphology of Abm-loaded nanoparticles for controlling *Meloidogyne incognita* were discussed. Notably, spindle-like nanoparticles were obtained with PCL-*b*-PEG as the stabilizer and found significantly more efficient (98.4% mortality at 1 ppm particle concentration) than spherical nanoparticles using PLGA-*b*-PEG or PLA-*b*-PEG as the stabilizer. This work provides a more rapid and powerful method to prepare stable Abm-loaded nanoparticles with tunable morphologies and improved effectiveness for controlling *Meloidogyne incognita*.

Keywords: *Meloidogyne incognita*; Abamectin; flash nanoprecipitation; amphiphilic block copolymers; spindle-like nanoparticles

1. Introduction

Plant-parasitic nematodes are one of the major agricultural pests worldwide, which have caused in excess of $157 billion in global crops damage annually [1]. Nematodes show a wide variety of interactions with their hosts, among which parasitic worms never enter the host and simply migrate through the soil, using roots as an ephemeral food source [2]. In particular, the root-knot nematodes (*Meloidogyne incognita*) are biotrophic parasite that draw nutrition from the root of hosts with a rich food source for several weeks, causing premature death of the host and reducing crop yields [3].

Traditionally, plants are grown in the soil in which *Meloidogyne incognita* have been controlled through the use of synthetic nematicides, such as fumigant nematicides [4]. Nevertheless, the use of these types of pesticides is undesirable due to problems of residual toxicity, environmental pollution, and public health hazards, etc. Alternative non-fumigant nematicides have been used [5], but they are inefficient in the control of root-knot nematodes because of their rapid degradation by soil microorganisms after repetitive use [6]. One biological and high-efficiency non-fumigant nematicide, Abamectin (Abm), has been widely used to control *Meloidogyne incognita* since the early 1980s [7]. However, one basic and formidable problem of Abm is its low solubility in water. In many cases, it is applied in the form of oil in water emulsion [8], which shows limited steady shelf time and binds to organic contents in crops. Another drawback of Abm is its degradation by photo-oxidation [9]. As a result, in practical application, most of the applied Abm is lost because of degradation under UV irradiation. Hence, Abm's dose is always increased to ensure efficacy, which results in increased costs and environmental pollution [10,11].

Nowadays, the rapid development of nanotechnology presents a new way to improve the performances of conventional pesticide formulations through the construction of nanotechnology-based agricultural systems such as the drug-carrier and a controllable drug targeting and releasing system [12–20]. Recently, there have been many types of nanomaterials for these nanotechnology-based systems, such as Ag-based nanocomposite structure for antibacterial [21], MoS_2 nanosheets nanostructures for targeted chemotherapy [22], and amphiphilic block copolymers (BCPs) for stabilizing drugs. Generally, drug molecules can be encapsulated inside the formed nanoparticles by using biocompatible nanomaterial BCPs as the carrier and stabilizer to enhance the drug stability and bioavailability. The resulted drug-loaded nanoparticles are able to protect the Abm against degradation, control the release rate, and prolong the duration of efficacy of Abm pesticide formulations [23–25].

Flash nanoprecipitation (FNP) is a simple and generic method to rapidly construct nanosized drug-loaded particles by copolymer-directed assembly [26,27]. FNP involves rapid micromixing of organic solutions of the hydrophobic drug and BCP with water (anti-solvent) in a multi-inlet vortex mixer (MIVM) to create high supersaturations of the drug in milliseconds, and then rapidly form the hydrophobic core (drug) in the mixed solvent. These hydrophobic cores are subsequently stabilized and protected from aggregation by the BCP [28–31], as illustrated in Figure 1a. The FNP method has been demonstrated to be powerful for the preparation of drug-loaded nanoparticles with high drug-loading capacity, relatively narrow size distribution, and tunable nanometer particle size [32–34]. In FNP, the drug-loaded nanoparticles with spherical morphology using BCP as the stabilizer have been extensively studied both experimentally and theoretically. However, drug-loaded particles with non-spherical nanostructure obtained by FNP methods are seldom reported, although these particles have attracted increasing attention in biological field thanks to their unique features and properties, including longer blood circulation time, better motions under flow conditions, and stronger adhesiveness to the biological substrate [35,36]. Therefore, further study into FNP for a preparation method capable of controlling the morphology of drug-loaded nanoparticles is still needed.

In this study, Abm-loaded nanoparticles suspensions with excellent stability were generated via FNP. Three biocompatible BCPs, poly(lactic-co-glycolic acid)-*b*-poly(ethylene glycol) (PLGA-*b*-PEG), poly(D,L-lactide)-*b*-poly(ethylene glycol) (PLA-*b*-PEG), and poly(caprolactone)-*b*-poly(ethylene glycol) (PCL-*b*-PEG) with a molecular weight of 10k-*b*-5k were chosen as the stabilizers. The Abm-to-stabilizer feed ratio was optimized, and Abamectin loading capacity and encapsulation efficiency were evaluated. Spherical and spindle-like morphologies were observed for the obtained Abm-loaded nanoparticles (NPs) by FNP (FNP-NPs). The toxicity of Abm-loaded nanoparticles with these two morphologies on *Meloidogyne incognita* was investigated.

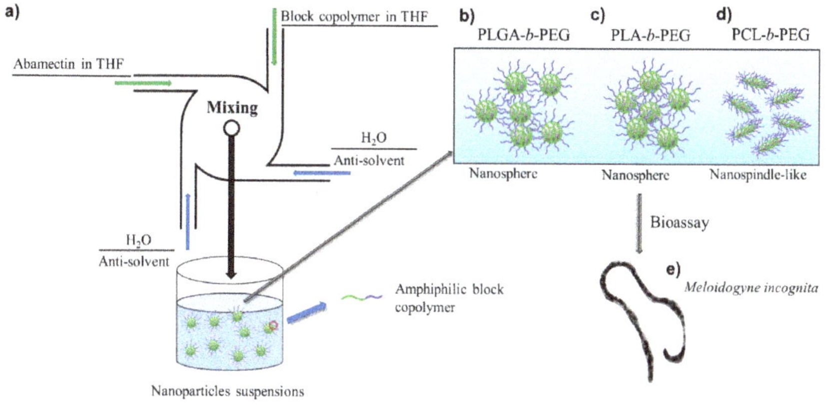

Figure 1. (**a**) Illustration of the preparation of Abamectin (Abm)-loaded nanoparticles by flash nanoprecipitation. (**b–d**) Morphology of Abm-loaded nanoparticles with different stabilizers: poly(lactic-co-glycolic acid)-*b*-poly(ethylene glycol) (PLGA-*b*-PEG) (**b**), poly(D,L-lactide)-*b*-poly(ethylene glycol) (PLA-*b*-PEG) (**c**), and poly(caprolactone)-*b*-poly(ethylene glycol) (PCL-*b*-PEG) (**d**). (**e**) Biological assay of Abm-loaded nanoparticles to *Meloidogyne incognita*. THF = tetrahydrofuran.

2. Materials and Methods

2.1. Materials

Abamectin (95.5%) was supplied by Hebei Veyong Bio-Chemical Co., Ltd. (Shijiazhuang, China). Poly(lactic-co-glycolic acid)-*b*-poly(ethylene glycol) (PLGA-*b*-PEG, 10k-*b*-5k) and poly(D,L-lactide)-*b*-poly(ethylene glycol) (PLA-*b*-PEG, 10k-*b*-5k) were purchased from Jinan Daigang Biomaterial Co., Ltd. (Jinan, China); ε-Caprolactone (ε-CL), stannous octoate, anhydrous toluene, and monomethoxy poly(ethylene glycol) (MW = 5000) were purchased from Sigma-Aldrich; diethyl ether was purchased from Shanghai Tianlian Fine Chemical Co., Ltd. (Shanghai, China), and tetrahydrofuran (THF) was purchased from Beijing Chemical Reagents Company (Beijing, China). Poly(caprolactone)-*b*-poly(ethylene glycol) (PCL-*b*-PEG, 10k-*b*-5k) was synthesized according to a previously reported method [32].

A population of root-knot nematodes (*Meloidogyne incognita*) was originally obtained from a greenhouse in the Beijing Academy of Agriculture and Forestry, Beijing, China.

2.2. Preparation of Abm-Loaded Nanoparticles

Abm-loaded nanoparticles were prepared by FNP method using a four-stream MIVM (Figure 1a). The amphiphilic block copolymer (PLGA-*b*-PEG, PLA-*b*-PEG, or PCL-*b*-PEG; 10k-*b*-5k) and Abamectin were dissolved in THF. Different drug-to-stabilizer feed ratios (weight) were used: 1:10, 2.5:10, 7.5:10, and 10:10. Solutions of Abm and BCP in THF (streams 1 and 2, 25 °C) were fed at the same flow rate (12 mL/min) along with two other deionized water streams (streams 3 and 4, 25 °C) both at a flow rate of 24 mL/min into the MIVM. The corresponding Reynolds number (*Re*) was calculated to be 5962, which is within the turbulent flow region with better mixing. The concentrations of Abm were 1, 2.5, 7.5, or 10 mg/mL, and the BCP concentration was fixed at 10 mg/mL. It is important to note that the concentrations of Abm and BCP in final nanoparticle solution was decreased due to the merging of the four streams. In addition, the morphology of FNP-NPs was strongly influenced by the glass transition temperature (T_g) of the BCP used, which affects the assembly of BCP on the hydrophobic core (drug) during FNP, and a low T_g could result in non-spherical nanoparticles under the millisecond mixing conditions of FNP [33]. In order to investigate the potential of FNP to form Abm-loaded nanoparticles with non-spherical morphologies, we used different BCPs with an appropriate hydrophilic to hydrophobic block molecular weight ratio.

2.3. Characterization

Particle size and size distribution were measured by dynamic light scattering (DLS) with a ZetasizerNano ZS90 (Malvern Instruments, Malvern, UK). The light intensity correlation function was collected at 25 °C with a scattering angle of 90°. The values obtained were averaged from three duplicates. Nanoparticle morphology was observed by transmission electron microscopy (TEM) (Hitachi HT7700, Hitachi Ltd., Chiyoda-ku, Japan) with an acceleration voltage of 80 kV. The samples were prepared by dripping the fresh solution onto carbon-coated copper grids and then dried overnight at room temperature. The amount of drug loaded in nanoparticles was determined by drug absorption at 245 nm using high performance liquid chromatography (HPLC) (Aglient 1260, Santa Clara, CA, USA) with a C18 column (5 μm, 4.6 mm × 150 mm, Aglient, Santa Clara, CA, USA) and a 245-nm UV detector.

2.4. Drug Loading Capacity (DLC) and Encapsulation Efficiency (EE)

To determine the DLC and EE of Abm in Abm-loaded nanoparticles, the nanoparticle solutions with a known volume (~5 mL) were dialyzed for 24 h using a dialysis bag (molecular weight cut off: 8-14 kDa), and the Milli-Q water was changed six times to remove the organic solvent down to a barely detectable level (i.e., <0.008 v/v %). For each change of Milli-Q water, the volume of water in the dialysis bag was slightly increased. After removal of the free Abm and residual organic solvent by dialysis, Abm concentration was examined by HPLC at 245 nm. The solution was diluted by methanol and filtered through a 0.2-μm filter prior to HPLC analysis. Then, the DLC and EE of Abm-loaded nanoparticles were calculated according to the following equations:

$$\mathrm{DLC}(\%) = \frac{\text{Total mass of loaded Abamectin}}{\text{Total mass of nanoparticles}} \times 100$$

$$\mathrm{EE}(\%) = \frac{\text{Total amount of loaded Abamectin}}{\text{Total amount of Abamectin added}} \times 100$$

The samples after dialysis were packed in glass tubes which were then stored at 0 °C for 7 days or 54 °C for 14 days, and the changes in DLC in the Abm-loaded nanoparticles were studied.

2.5. Biological Assay

All juveniles were hatched in a sieve (mesh number 500) at 28 °C for 3 days, and then juveniles were collected and used in the following experiments.

The nematicidal efficacy of Abamectin against *Meloidogyne incognita* was determined in aqueous tests. Abamectin nanoparticle suspensions formulations (1, 0.5, 0.25, 0.125, and 0.0625 mg/L) were prepared by diluting the nanoparticle solution directly obtained from FNP with sterile water, and pure sterile water was used as the control. Then, 0.5 mL of the resulting solution and 0.5 mL of root-knot nematodes juveniles (containing an average of 50 juveniles) were added to each well of a 24-well plate, respectively. Well plates were wrapped and kept at 28 °C. After 24 h, the relative percentages of the motile and immotile juveniles were evaluated by observation under a microscope (Olympus, Tokyo, Japan) three times. Each experiment had four duplicates and the final results were obtained by averaging.

3. Results and Discussion

3.1. Effect of BCP on Size and Morphology of Abm-Loaded Particles

An important finding associated with BCP is that they have the potential to control the particle morphology and size when they assemble on the particle core (drug). Most of the Abm-loaded particles reported have spherical morphology, which is most likely to occur [23,37,38]. For many applications, however, non-spherical structures are desired, making it important to control the particle morphology.

Focusing on preparing non-spherical particles, we found that the particle morphology can be controlled by using different BCPs as the stabilizer in FNP. Figure 2a,b show the typical TEM images of the prepared Abm-loaded nanoparticles using PLGA-*b*-PEG and PLA-*b*-PEG as stabilizers by FNP, respectively. The morphology of these particles is perfectly spherical and the particle size distribution is narrow, which is favorable for improving the dispersion, adhesion, and permeability of pesticides to target crops as a pesticide carrier [39]. Interestingly, as the BCP was changed to PCL-*b*-PEG, spindle-like particles were obtained (Figure 2c). The resulting nanoparticles have a long axis length ranging in 190–210 nm and a short axis length ranging in 70–80 nm instead of a normal spherical structure. Our observation of spindle-like Abm-loaded particles is an important finding because it is the first demonstration that the non-spherical drug-loaded particles are produced through FNP method. This finding also confirms that the spindle-like shape of PCL-*b*-PEG stabilized Abm-loaded nanoparticles due to the lower T_g of the hydrophobic PCL block, which causes the larger degree of stretching of the poly(caprolactone) (PCL) block. Meanwhile, the degree of stretching of the PCL block is too high to support the spherical morphology [40], leading to spindle-like morphology of the FNP-NPs. In addition, it is noteworthy that the edges of the nanoparticles were not smooth. This may be due to the presence of poly(ethylene glycol) (PEG) blocks which could affect the folding of the PCL chain [41]. Changes in drug-loaded particle shape could significantly change the drug release behavior, as suggested in the literature [42]. Thus, the relevance of spherical and non-spherical morphologies of Abm-loaded FNP-NPs and their nematicidal activity will be discussed in detail in a later section.

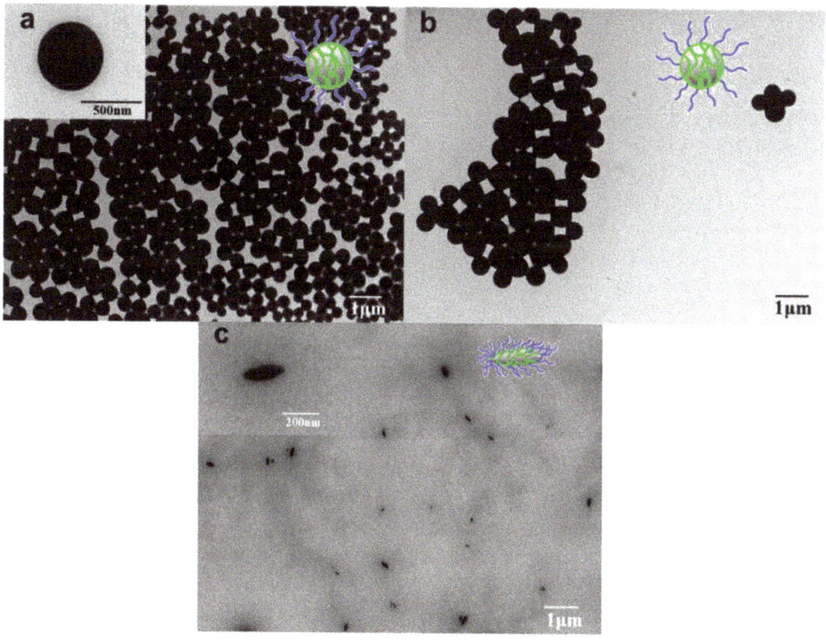

Figure 2. TEM photographs of Abm-loaded nanoparticles prepared using (**a**) PLGA-*b*-PEG, (**b**) PLA-*b*-PEG, and (**c**) PCL-*b*-PEG as the stabilizer, respectively. The insets are the corresponding schematic diagrams.

Figure 3 displays the average size (diameter) and size distribution of the prepared Abm-loaded particles when different BCPs were used as the stabilizer. It can be seen that the three kinds of FNP-NPs all had relatively narrow size distribution, validating the superiority of FNP. Meanwhile, it can be seen that PCL-*b*-PEG stabilized nanoparticles have smaller average sizes (72 nm) than those stabilized by PLGA-*b*-PEG or PLA-*b*-PEG (with an average size 414 or 314 nm). The most likely reason is that PCL, the hydrophobic block, has a much lower T_g (about −60 °C), making the BCP chain softer. Consequently, the drug is wrapped very rapidly, hindering the further growth of the drug nuclei and resulting in a smaller particle size. In contrast, PLGA-*b*-PEG and PLA-*b*-PEG have hydrophobic blocks with much higher T_g values (about 39 °C and 34 °C for poly(lactic-co-glycolic acid) (PLGA) and poly(D,L-lactide) (PLA), respectively) [43,44]. Thus the chains of these two BCPs are not flexible enough to encapsulate the drug nuclei in time before it grows larger, leading to much larger particle sizes.

Meanwhile, as suggested in the literature, smaller sizes (with large surface-volume ratios) could provide better performance, such as easy attachment and unique optical properties [45]. Therefore, the small spindle-like Abm-loaded nanoparticles obtained in our work could have higher potential in the control of *Meloidogyne incognita* (which will be validated in a later section).

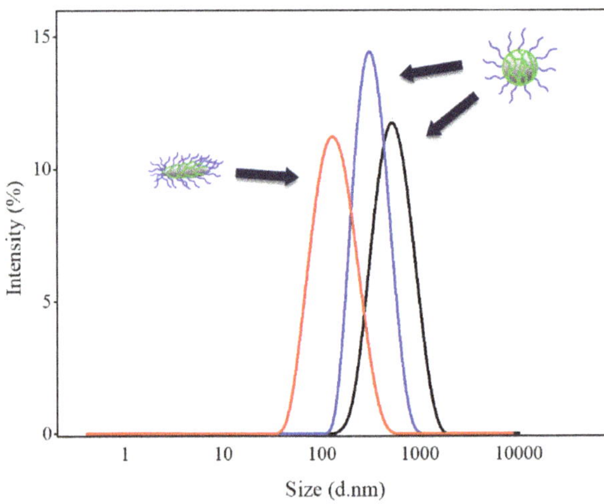

Figure 3. Particle size and size distribution of Abm-loaded nanoparticles prepared using PLGA-*b*-PEG (**black**), PLA-*b*-PEG (**blue**), and PCL-*b*-PEG (**red**) as the stabilizer, respectively. The mean diameters derived from the Gaussian fits (solid lines) are 414, 314, and 72 nm, respectively.

3.2. *Effect of Abm-to-Stabilizer Feed Ratio on Particle Stability*

The dependence of particle size on feed ratio (weight) of Abm-to-stabilizer was exploited to optimize the particle stability which could be indicated by the change of particle size over time (the change of drug loading capacity along with time will be discussed in a later section). Abm-loaded nanoparticles prepared with different Abm-to-stabilizer feed ratios and with PLGA-*b*-PEG, PLA-*b*-PEG, or PCL-*b*-PEG as the stabilizer were stored at room temperature, and their particle size was monitored at defined time intervals. As shown in Figure 4 (with PLGA-*b*-PEG as the stabilizer), a feature that we saw for all the nanoparticles in the first three days was the "anti-Ostwald" phenomenon, which may be related to polar hydroxyl of Abm in nanoparticles. This polar moiety slowly rearranges towards the particle interface over time to minimize the energy of the system by increasing the surface to volume area, resulting in a decrease in particle size [46].

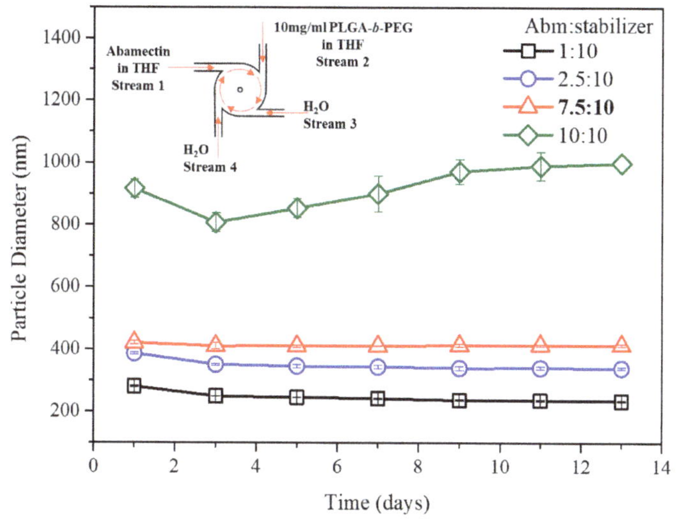

Figure 4. Effect of various Abm-to-stabilizer ratios on particle stability for flash nanoprecipitation-nanoparticles (FNP-NPs) prepared with 10 mg/mL PLGA-b-PEG as the stabilizer. Stream 1 was 1, 2.5, 7.5, or 10 mg/mL of Abamectin dissolved in THF. Stream 2 was 10 mg/mL of PLGA-*b*-PEG dissolved in THF. The other two streams were both water.

After the first three days, a normal feature that we saw for nanoparticles with a feed ratio (weight) of 10:10 (Abm/PLGA-*b*-PEG) was "Ostwald ripening" [46,47]. This is because the amount of stabilizer was not enough to encapsulate the drug nuclei in time to prevent their further growth with the high drug-to-stabilizer feed ratio 10:10, leading to the relatively wide size distribution of the initially obtained particles (polydispersity index (PDI) = 0.53, Table 1). Consequently, small nanoparticles would continuously dissolve to precipitate again on the surface of larger nanoparticles, and the average particle size increased gradually as shown in Figure 4. In contrast, particle stability was considerably enhanced when the Abm-to-stabilizer ratio was reduced to 7.5:10, 2.5:10, or 1:10, and the particle size remained essentially unchanged 10 days later. In addition, Abm loading increased with an increase in Abm-to-stabilizer feed ratio. This means the two lower Abm-to-stabilizer ratios (2.5:10 and 1:10) were inefficient at encapsulating Abm in nanoparticles. Hence, exhibiting remarkable stability and excellent drug loading capacity, the nanoparticles prepared with an Abm-to-stabilizer feed ratio of 7.5:10 (weight) were taken as the optimized formulation.

Table 1. Nanoparticle average size and size distribution with different Abm-to-stabilizer ratios (using PLGA-*b*-PEG as the stabilizer and the stabilizer concentration in THF before mixing was 10 mg/mL). PDI = polydispersity index.

Ratio of Abamectin to Stabilizer	Particle Diameter (nm)	PDI
1:10	252 ± 1	0.25 ± 0.01
2.5:10	355 ± 3	0.14 ± 0.04
7.5:10	414 ± 5	0.19 ± 0.07
10:10	898 ± 30	0.53 ± 0.20

3.3. The Amount of Abamectin Encapsulated in FNP-NPs

The solubility of Abamectin in water is extremely low, which results in precipitation of Abamectin. During FNP, almost all of the Abamectin (>99.9%) could be encapsulated and stabilized by the stabilizer, which means Abamectin molecules could be wrapped in nanoparticles efficiently with nearly no loss during dialysis [48].

Thus, high drug loading capacity is one of the advantages of FNP. The amounts of Abamectin in the FNP-NPs (DLC and EE) using different stabilizers, are summarized in Table 2. The concentration of Abamectin in the final nanoparticle solution reached around 0.8 mg/mL after dialysis. The encapsulation efficiency was all higher than 95% using the three different stabilizers (Table 2). The slight loss may be caused by imperfect operation in the mixer MIVM, in which fluids may have hold-up volume. The values of DLC of different particles stabilized by PLGA-*b*-PEG, PLA-*b*-PEG, and PCL-*b*-PEG were determined to be 41.46%, 40.97%, and 40.76%, respectively. Overall, high drug loading capacity and encapsulation efficiency were achieved by the FNP method for all stabilizers at room temperature.

Table 2. The amount of Abamectin encapsulated in Abm-NPs (Abm-to-stabilizer feed ratio 7.5:10). DLC = Drug Loading Capacity; EE = Encapsulation Efficiency.

	Stabilizer	DLC (%)	EE (%)
Abm-NPs-1	PLGA-*b*-PEG	41.46 ± 0.05	96.74 ± 0.12
Abm-NPs-2	PLA-*b*-PEG	40.97 ± 0.05	95.60 ± 0.11
Abm-NPs-3	PCL-*b*-PEG	40.76 ± 0.03	95.10 ± 0.07

3.4. *Effect of Temperature on Particle Stability*

The storage stability of Abm-loaded nanoparticles was studied by measuring Abm loading capacity at two different temperatures (0 °C and 54 °C). The results are shown in Figure 5 (with PLGA-*b*-PEG as the stabilizer). Pleasingly, the Abm-loaded nanoparticles remained stable with no major changes in the loading capacity during storage at 0 °C, although the loss of Abm (9.09%) was found after storage for 14 days at 54 °C. The reason is that Abamectin degrades much faster at higher temperatures. These results show that the Abm-loaded nanoparticles can be kept in a very stable state during long-time storage at relatively low temperatures.

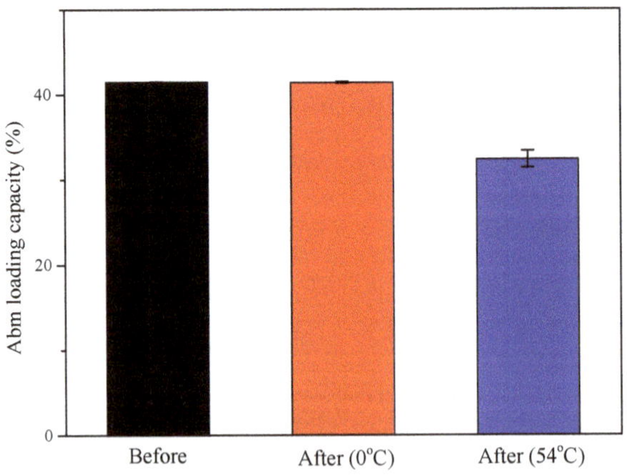

Figure 5. Abm loading capacity of the nanoparticles prepared using PLGA-b-PEG as the stabilizer before and after storage at 0 °C for 7 days and 54 °C for 14 days.

3.5. *Toxicity of Abm-Loaded Particles to Meloidogyne incognita*

As shown in Figure 6, all Abm-loaded nanoparticles still cause high mortality (87.2%, 97.7%, and 98.4% with PLGA-*b*-PEG, PLA-*b*-PEG, and PCL-*b*-PEG as the stabilizer, respectively) of

Meloidogyne incognita even after being diluted 800 times (800 ppm to 1 ppm). The lethal concentrations required to kill 50% (LC_{50}) (after 24 h) for *Meloidogyne incognita* were calculated and are shown in Table 3. The values of LC_{50} were 0.42, 0.37, and 0.28 ppm with PLGA-*b*-PEG, PLA-*b*-PEG, and PCL-*b*-PEG as the stabilizer, respectively (Table 3). The mean diameters of the three nanoparticles are 414, 314, and 72 nm, respectively (Figure 3), and they have the same Abm concentrations (0.8 mg/mL), but different LC_{50} values. One possible reason is that nanoparticles formed from BCPs containing a hydrophobic block (PLGA or PLA) with a high T_g tend to disassemble more slowly than those formed from BCPs with a low T_g hydrophobic block (PCL) [49]. Another possible reason is that the spindle-like nanoparticles obtained from PCL-*b*-PEG with smaller sizes have better adhesion and permeability.

Overall, the lethal concentration assays indicate that spindle-like Abm-loaded nanoparticles prepared using PCL-*b*-PEG as the stabilizer by FNP have higher nematicidal efficiency than spherical Abm-loaded nanoparticles using PLGA-*b*-PEG or PLA-*b*-PEG as the stabilizer. In addition, Figure 6 shows that the Abm-loaded nanoparticles caused irreversible paralysis in *Meloidogyne incognita*, and the nematode mortality increased with the increasing concentration of Abm-loaded nanoparticles. Thus, the nematode control effectiveness can be improved by using increased concentration of Abm-loaded nanoparticles [50].

Table 3. Bioassay results of the three Abamectin formulations. LC_{50} = lethal concentrations required to kill 50%. CL = confidence limit. DF = degree freedom. P = probability. n = number.

Population	Insecticide	Stabilizer	LC_{50} (95% CL [a]) (ppm)	Fit of Probit Line				
				Slope ± SE	χ^2	DF	P	n [b]
Lab	Abm-NPs-1	PLGA-*b*-PEG	0.42 (0.32–0.62)	5.76 ± 0.33	5.95	3	0.11	252
Lab	Abm-NPs-2	PLA-*b*-PEG	0.37 (0.32–0.48)	6.72 ± 0.59	0.79	3	0.84	320
Lab	Abm-NPs-3	PCL-*b*-PEG	0.28 (0.23–0.33)	6.57 ± 0.33	3.45	3	0.32	352

[a] 95% confidence limit. [b] Number of larvae used in the bioassay.

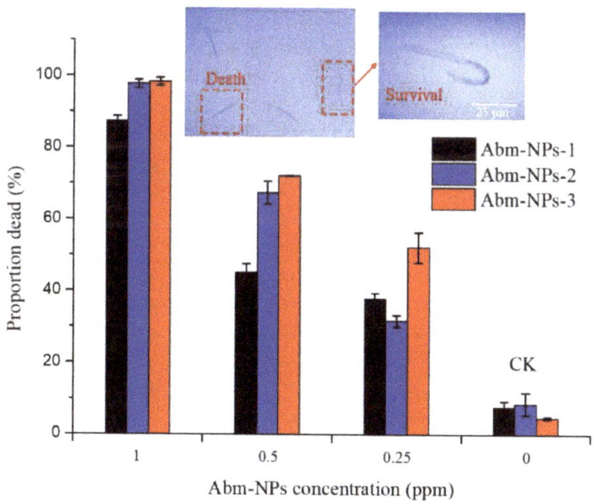

Figure 6. Mortality of *Meloidogyne incognita* as a function of concentration of Abm-loaded nanoparticles prepared using different block copolymers (BCPs) as the stabilizer.

Author Contributions: Z.F., K.C., L.L., F.Z., Y.W., M.W., Y.S., H.C., D.L., and X.G. were involved in the experimental works. Z.F. performed all the experiments and wrote the paper. K.C., M.W., Y.W., Y.S., H.C., and X.G. supervised some research steps and allowed supply of some of the reagents and access to some of the facilities necessary to perform experiments. L.L. and F.Z. revised and proof read the whole manuscript.

Acknowledgments: We gratefully thank the National Natural Science Foundation of China (21476143, 51773061 and 5171101370), 111 Project Grant (B08021), Open Project of Key Laboratory of Materials-Oriented Chemical Engineering of Xinjiang Uygur Autonomous Region (No. 2017BTRC002), Major Science and Technology Program for Water Pollution Control and Treatment (No. 2017ZX07101-003), and Major National Scientific Research Program of China (No. 2014CB932200) for financial support.

Conflicts of Interest: The authors declare no conflicts of interest.

References

1. Abad, P.; Gouzy, J.; Aury, J.M.; Castagnone-Sereno, P.; Danchin, E.G.J.; Deleury, E.; Perfus-Barbeoch, L.; Anthouard, V.; Artiguenave, F.; Blok, V.C.; et al. Genome sequence of the metazoan plant-parasitic nematode Meloidogyne incognita. *Nat. Biotechnol.* **2008**, *26*, 909–915. [CrossRef] [PubMed]
2. Jones, J.T.; Haegeman, A.; Danchin, E.G.J.; Gaur, H.S.; Helder, J.; Jones, M.G.K.; Kikuchi, T.; Manzanilla-López, R.; Palomares-Rius, J.E.; Wesemael, W.M.L.; et al. Top 10 plant-parasitic nematodes in molecular plant pathology. *Mol. Plant Pathol.* **2013**, *14*, 946–961. [CrossRef] [PubMed]
3. Jaouannet, M.; Magliano, M.; Arguel, M.J.; Gourgues, M.; Evangelisti, E.; Abad, P.; Rosso, M.N. The root-knot nematode calreticulin Mi-CRT is a key effector in plant defense suppression. *Mol. Plant Microbe Interact.* **2013**, *26*, 97–105. [CrossRef] [PubMed]
4. Desaeger, J.A.; Seebold, K.W.; Csinos, A.S. Effect of application timing and method on efficacy and phytotoxicity of 1,3-D, chloropicrin and metam-sodium combinations in squash plasticulture. *Pest Manag. Sci.* **2010**, *64*, 230–238. [CrossRef] [PubMed]
5. Stirling, A.M.; Stirling, G.R.; Macrae, I.C. Microbial degradation of Fenamiphos after repeated application to a tomato-growing soil. *Nematologica* **1992**, *38*, 245–254. [CrossRef]
6. Giannakou, I.O.; Karpouzas, D.G. Evaluation of chemical and integrated strategies as alternatives to methyl bromide for the control of root-knot nematodes in Greece. *Pest Manag. Sci.* **2003**, *59*, 883–892. [CrossRef] [PubMed]
7. Sasser, J.N.; Kirkpatrick, T.L.; Dybas, R.A. Efficacy of avermectins for root-knot control in tobacco. *Plant Dis.* **1982**, *66*, 691–693. [CrossRef]
8. Shi, Y.; Zheng, T.; Shang, Q. Preparation of acrylic/acrylate copolymeric surfactants by emulsion polymerization used in pesticide oil-in-water emulsions. *J. Appl. Polym. Sci.* **2012**, *123*, 3117–3127. [CrossRef]
9. Guimaraes, J.R.; Barbosa, I.M.; Maniero, M.G.; Rath, S. Abamectin degradation by Advanced Oxidation Processes: Evaluation of toxicity reduction using Daphnia similis. *J. Adv. Oxid. Tech.* **2014**, *17*, 82–92.
10. Erzen, N.K.; Kolar, L.; Flajs, V.C.; Kužner, J.; Marc, I.; Pogačnik, M. Degradation of abamectin and doramectin on sheep grazed pasture. *Ecotoxicology* **2005**, *14*, 627–635. [CrossRef] [PubMed]
11. Ali, S.W.; Li, R.; Zhou, W.Y.; Sun, J.Q.; Guo, P.; Ma, J.P.; Li, S.P. Isolation and characterization of an abamectin-degrading Burkholderiacepacia-like GB-01 strain. *Biodegradation* **2010**, *21*, 441–452. [CrossRef] [PubMed]
12. Khot, L.R.; Sankaran, S.; Maja, J.M.; Ehsania, R.; Schusterb, E.W. Applications of nanomaterials in agricultural production and crop protection: A review. *Crop Prot.* **2012**, *35*, 64–70. [CrossRef]
13. Mattos, B.D.; Tardy, B.L.; Magalhães, W.L.E.; Rojas, O.J. Controlled release for crop and wood protection: Recent progress toward sustainable and safe nanostructured biocidal systems. *J. Control. Release* **2017**, *262*, 139–150. [CrossRef] [PubMed]
14. Iavicoli, I.; Leso, V.; Beezhold, D.H.; Shvedova, A.A. Nanotechnology in agriculture: Opportunities, toxicological implications, and occupational risks. *Toxicol. Appl. Pharmacol.* **2017**, *329*, 96–111. [CrossRef] [PubMed]
15. De Oliveira, J.L.; Campos, E.V.; Bakshi, M.; Abhilash, P.C.; Fraceto, L.F. Application of nanotechnology for the encapsulation of botanical insecticides for sustainable agriculture: Prospects and promises. *Biotechnol. Adv.* **2014**, *32*, 1550–1561. [CrossRef] [PubMed]
16. Wang, Y.; Wang, A.; Wang, C.; Cui, B.; Sun, C.; Zhao, X.; Zeng, Z.; Shen, Y.; Gao, F.; Liu, G.; et al. Synthesis and characterization of emamectin-benzoate slow-release microspheres with different surfactants. *Sci. Rep.* **2017**, *7*, 12761. [CrossRef] [PubMed]
17. Shen, Y.; Wang, Y.; Zhao, X.; Sun, C.; Cui, B.; Gao, F.; Zeng, Z.; Cui, H. Preparation and Physicochemical Characteristics of Thermo-Responsive Emamectin Benzoate Microcapsules. *Polymers* **2017**, *9*, 418. [CrossRef]

18. Liu, B.; Wang, Y.; Yang, F.; Wang, X.; Shen, H.; Cui, H.; Wu, D. Construction of a controlled-release delivery system for pesticides using biodegradable PLA-based microcapsules. *Colloids Surf. B* **2016**, *144*, 38–45. [CrossRef] [PubMed]
19. Yamamoto, E.; Kuroda, K. Colloidal Mesoporous Silica Nanoparticles. *Bull. Chem. Soc. Jpn.* **2016**, *89*, 501–539. [CrossRef]
20. Nakamura, M.; Tahara, Y.; Fukata, S.; Zhang, M.; Yang, M.; Iijima, S.; Yudasaka, M. Significance of Optimization of Phospholipid Poly(Ethylene Glycol) Quantity for Coating Carbon Nanohorns to Achieve Low Cytotoxicity. *Bull. Chem. Soc. Jpn.* **2017**, *90*, 662–666. [CrossRef]
21. Tang, Q.; Liu, J.; Shrestha, L.K.; Ariga, K.; Ji, Q. Antibacterial Effect of Silver-Incorporated Flake-Shell Nanoparticles under Dual-Modality. *ACS Appl. Mater. Interfaces* **2016**, *8*, 18922–18929. [CrossRef] [PubMed]
22. Li, B.L.; Setyawati, M.I.; Chen, L.; Xie, J.; Ariga, K.; Lim, C.-T.; Garaj, S.; Leong, D.T. Directing Assembly and Disassembly of 2D MoS_2 Nanosheets with DNA for Drug Delivery. *ACS Appl. Mater. Interfaces* **2017**, *9*, 15286–15296. [CrossRef] [PubMed]
23. Wang, Y.; Cui, H.; Sun, C.; Zhao, X.; Cui, B. Construction and evaluation of controlled-release delivery system of Abamectin using porous silica nanoparticles as carriers. *Nanoscale Res. Lett.* **2014**, *9*, 655. [CrossRef] [PubMed]
24. Yu, M.; Yao, J.; Liang, J.; Zeng, Z.; Cui, B.; Zhao, X.; Sun, C.; Wang, Y.; Liu, G.; Cui, H. Development of functionalized abamectinpoly (lactic acid) nanoparticles with regulatable adhesion to enhance foliar retention. *RSC Adv.* **2017**, *7*, 11271–11280. [CrossRef]
25. Cao, J.; Guenther, R.H.; Sit, T.L.; Lommel, S.A.; Opperman, C.H.; Willoughby, J.A. Development of Abamectin loaded plant virus nanoparticles for efficacious plant parasitic nematode control. *ACS Appl. Mater. Interfaces* **2015**, *7*, 9546–9553. [CrossRef] [PubMed]
26. Wang, M.; Xu, Y.; Wang, J.; Liu, M.; Yuan, Z.; Chen, K.; Li, L.; Prud'Homme, R.K.; Guo, X. Biocompatible nanoparticle based on dextran-*b*-poly(L-Lactide) block copolymer formed by flash nanoprecipitation. *Chem. Lett.* **2015**, *44*, 1688–1690. [CrossRef]
27. Wang, M.; Yang, N.; Guo, Z.; Gu, K.; Shao, A.; Zhu, W.; Xu, Y.; Wang, J.; Prud'Homme, R.K.; Guo, X. Facile Preparation of AIE-active fluorescent nanoparticles through flash nanoprecipitation. *Ind. Eng. Chem. Res.* **2015**, *54*, 4683–4688. [CrossRef]
28. Johnson, B.K.; Prud'Homme, R.K. Mechanism for rapid self-assembly of block copolymer nanoparticles. *Phys. Rev. Lett.* **2003**, *91*, 118302. [CrossRef] [PubMed]
29. Saad, W.S.; Prud'Homme, R.K. Principles of nanoparticle formation by Flash Nanoprecipitation. *Nano Today* **2016**, *11*, 212–227. [CrossRef]
30. Liu, Y.; Cheng, C.; Liu, Y.; Prud'Homme, R.K.; Fox, R.O. Mixing in a multi-inlet vortex mixer (MIVM) for flash nano-precipitation. *Chem. Eng. Sci.* **2008**, *63*, 2829–2842. [CrossRef]
31. Russ, R.; Liu, Y.; Prud'homme, R.K. Optimized descriptive model for micromixing in avortexmixer. *Chem. Eng. Commun.* **2010**, *197*, 1068–1075. [CrossRef]
32. Fu, Z.; Li, L.; Wang, M.; Guo, X. Size control of drug nanoparticles stabilized by mPEG-b-PCL during flash nanoprecipitation. *Colloid Polym. Sci.* **2018**, *296*, 935–940. [CrossRef]
33. Zhu, Z. Effects of amphiphilic diblock copolymer on drug nanoparticle formation and stability. *Biomaterials* **2013**, *34*, 10238–10248. [CrossRef] [PubMed]
34. Pustulka, K.M.; Wohl, A.R.; Lee, H.S.; Michel, A.R.; Han, J.; Hoye, T.R.; McCormick, A.V.; Panyam, J.; Macosko, C.W. Flash nanoprecipitation: Particle structure and stability. *Mol. Pharm.* **2013**, *10*, 4367–4377. [CrossRef] [PubMed]
35. Tuncelli, G.; Ay, A.N.; Zümreoglu-Karan, B. 5-Fluorouracil intercalated iron oxide@layered double hydroxide core-shell nano-composites with isotropic and anisotropic architectures for shape-selective drug delivery applications. *Mater. Sci. Eng. C* **2015**, *55*, 562–568. [CrossRef] [PubMed]
36. Geng, Y.; Dalhaimer, P.; Cai, S.; Tsai, R.; Tewari, M.; Minko, T.; Discher, D.E. Shape effects of filaments versus spherical particles in flow and drug delivery. *Nat. Nanotechnol.* **2007**, *2*, 249–255. [CrossRef] [PubMed]
37. Liu, Z.; Qie, R.; Li, W.; Hong, N.; Li, Y.; Li, C.; Wang, R.; Shi, Y.; Guo, X.; Jia, X. Preparation of avermectin microcapsules with anti-photodegradation and slow-release by the assembly of lignin derivatives. *New J. Chem.* **2017**, *41*, 3190–3195. [CrossRef]

38. Li, D.; Liu, B.; Yang, F.; Wang, X.; Shen, H.; Wu, D. Preparation of uniform starch microcapsules by premix membrane emulsion for controlled release of avermectin. *Carbohydr. Polym.* **2016**, *136*, 341–349. [CrossRef] [PubMed]
39. Zhao, X.; Cui, H.; Wang, Y.; Sun, C.; Cui, B.; Zeng, Z. Development strategies and prospects of nano-based smart pesticide formulation. *J. Agric. Food Chem.* **2017**. [CrossRef] [PubMed]
40. Chen, W.H.; Hua, M.Y.; Lee, R.S. Synthesis and characterization of poly(ethylene glycol)-*b*-poly(ε-caprolactone) copolymers with functional side groups on the polyester block. *J. Appl. Polym. Sci.* **2012**, *125*, 2902–2913. [CrossRef]
41. Wang, J.; Zhu, W.; Peng, B.; Chen, Y. A facile way to prepare crystalline platelets of block copolymers by crystallization-driven self-assembly. *Polymer* **2013**, *54*, 6760–6767. [CrossRef]
42. Rytting, E.; Nguyen, J.; Wang, X.; Kissel, T. Biodegradable polymeric nanocarriers for pulmonary drug delivery. *Expert Opin. Drug Deliv.* **2008**, *5*, 629–639. [CrossRef] [PubMed]
43. Houchin, M.L.; Topp, E.M. Physical properties of PLGA films during polymer degradation. *J. Appl. Polym. Sci.* **2009**, *114*, 2848–2854. [CrossRef]
44. Omelczuk, M.O.; McGinity, J.W. The influence of polymer glass transition temperature and molecular weight on drug release from tablets containing poly(dl-lactic acid). *Pharm. Res.* **1992**, *9*, 26–32. [CrossRef] [PubMed]
45. Ghormade, V.; Deshpande, M.V.; Paknikar, K.M. Perspectives for nano-biotechnology enabled protection and nutrition of plants. *Biotechnol. Adv.* **2011**, *29*, 792–803. [CrossRef] [PubMed]
46. Kumar, V.; Adamson, D.H.; Prud'Homme, R.K. Fluorescent polymeric nanoparticles: Aggregation and phase behavior of pyrene and amphotericin B molecules in nanoparticle cores. *Small* **2010**, *6*, 2907–2914. [CrossRef] [PubMed]
47. D'Addio, S.M.; Prud'Homme, R.K. Controlling drug nanoparticle formation by rapid precipitation. *Adv. Drug Deliv. Rev.* **2011**, *63*, 417–426. [CrossRef] [PubMed]
48. Liu, Y.; Tong, Z.; Prud'Homme, R.K. Stabilized polymeric nanoparticles for controlled and efficient release of bifenthrin. *Pest Manag. Sci.* **2008**, *64*, 808–812. [CrossRef] [PubMed]
49. Allen, C.; Maysinger, D.; Eisenberg, A. Nano-engineering block copolymer aggregates for drug delivery. *Colloids Surf. B* **1999**, *16*, 3–27. [CrossRef]
50. Faske, T.R.; Starr, J.L. Sensitivity of Meloidogyne incognita and Rotylenchulus reniformis to Abamectin. *J. Nematol.* **2006**, *38*, 240–244. [PubMed]

nanomaterials

MDPI

Article

Removing Metal Ions from Water with Graphene–Bovine Serum Albumin Hybrid Membrane

Xiaoqing Yu [1,†], Shuwei Sun [1,†], Lin Zhou [2], Zhicong Miao [1], Xiaoyuan Zhang [3,*], Zhiqiang Su [1,*] and Gang Wei [4,*]

1 State Key Laboratory of Chemical Resource Engineering, Beijing Key Laboratory of Advanced Functional Polymer Composites, Beijing University of Chemical Technology, Beijing 100029, China; yu_xiaoq@163.com (X.Y.); 2017200438@mail.buct.edu.cn (S.S.); 18810867810@163.com (Z.M.)
2 School of Chemical Engineering and Technology, Tianjin University, Tianjin 300072, China; linzhou@tju.edu.cn
3 Otto Schott Institute of Materials Research, Friedrich Schiller University Jena, Löbdergraben 32, 07743 Jena, Germany
4 Faculty of Production Engineering, University of Bremen, D-28359 Bremen, Germany
* Correspondence: xiaoyuan.zhang@uni-jena.de (X.Z.); suzq@mail.buct.edu.cn (Z.S.); wei@uni-bremen.de (G.W.)
† These authors contributed equally to this work.

Received: 21 January 2019; Accepted: 13 February 2019; Published: 16 February 2019

Abstract: Here we report the fabrication of graphene oxide (GO)-based membranes covalently combined with bovine serum albumin (BSA) for metal ions detection. In this system, BSA acts as a transporter protein in the membrane and endows the membrane with selective recognition of Co^{2+}, Cu^{2+}, $AuCl_4^-$, and Fe^{2+}. Combining the metal-binding ability of BSA and the large surface area of GO, the hybrid membrane can be used as a water purification strategy to selectively absorb a large amount of $AuCl_4^-$ from $HAuCl_4$ solution. Moreover, BSA could reduce the membrane-immobilized $AuCl_4^-$ by adding sodium borohydride ($NaBH_4$). Interestingly, adsorption experiments on three kinds of metal ions showed that the GO–BSA membrane had good selective adsorption of Co^{2+} compared with Cu^{2+} and Fe^{2+}. The morphology and composition changes of the membrane were observed with atomic force microscopy (AFM) and Raman spectroscopy, respectively. It is expected that this facile strategy for fabricating large-scale graphene-biomolecule membranes will spark inspirations in the development of functional nanomaterials and wastewater purification.

Keywords: protein; self-assembly; graphene oxide; membrane; water purification

1. Introduction

Purification of wastewater is always the focus of industry, pharmaceutical enterprises, and environmental protection departments [1]. Wastewater contains not only dyes but also metal ions, such as Co^{2+}, Cu^{2+}, and Fe^{2+} [2]. In addition, the wastewater from gold purification contains residual $AuCl_4^-$, which causes severe healthcare problems for human beings [3]. Therefore, the separation of precious metal ions from wastewater is of great significance for wastewater purification, and to some extent, for metal recovery. Among various biomolecules, proteins have been suggested as a potential candidate for removing heavy metal ions from water [4–6]. For instance, Mezzenga et al. [7,8] reported that β-lactoglobulin amyloid fibril could be combined with activated carbon and used for the efficient removal of heavy metal ion from water. Their findings indicated that these protein fibrils allowed the reduction of membrane-immobilized metal ions into valuable metal nanoparticles. Another work showed that bovines serum albumin (BSA) can reduce Au(III) ions into Au clusters in the presence of NaOH [9].

As a typical single atomic two-dimensional nanomaterial, graphene exhibits a large specific surface area and high charge mobility, which endows it with applications in material science [10,11], nanotechnology [12], analytical science [13,14], biomedicine [15–17], wastewater purification [18–20], and other fields [21]. In particular, chemically-modified graphene nanomaterials, such as graphene oxide (GO) [22–24], reduced graphene oxide (RGO) [25–27], and graphene quantum dots (GQDs) [28, 29], can improve water dispersion [30–32]. Graphene oxide and RGO were reported to be capable of effective water restoration and adsorption of toxic gases [33]. On the one hand, the oxygen-containing functional groups on the surface of GO, such as carboxyl and hydroxyl groups, can adsorb metal cations to achieve water purification. On the other hand, the products left can be easily separated from water [34]. In order to selectively adsorb the pollutants in water [35], many GO hybrid nanomaterials have been prepared with molecules such as polymers [36], peptides [37], proteins [38], and other small molecules [39]. Superior to other molecules, proteins become excellent biomolecules to bind with GO because of their low cost and rich functional groups. Most recently, Lu et al. [10] applied β-lactoglobulin on the surface of RGO sheets. The composite was used as a template for gold nanoparticle (AuNP) assembly. These AuNPs assembled on the β-lactoglobulin–RGO composites yielded an improved surface-enhanced Raman scattering (SERS) for Rhodamine 6G (Rd6G). In other cases, BSA can also be combined with GO to efficiently detect and extract Hg^{2+} [40]. Therefore, fabricating GO-protein composite membranes is of great value for reducing metal ions to metal nanoparticles and further achieving the biomimetic membrane in sewage treatment [35].

Inspired by the above studies, we provide a facile method to fabricate the GO–BSA membrane for removing heavy metal ions from water. Firstly, the GO prepared by the Hammers method was mixed with $ClCH_2COONa$ and NaOH to transform the esters, hydroxyl groups, and epoxies into carboxyl groups on the surface of GO. Then we fabricated the GO–BSA membrane by incorporating BSA on GO–COOH via vacuum filter (Figure 1). The GO–BSA membrane was used to remove metal ions, i.e., $AuCl_4^-$, Co^{2+}, Fe^{2+}, and Cu^{2+} ions from water.

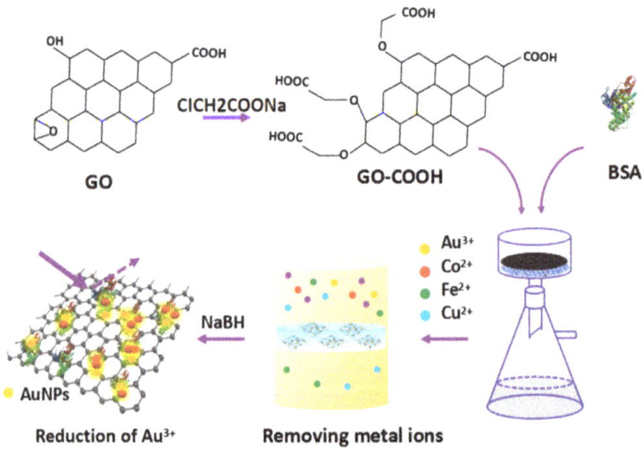

Figure 1. Schematic illustration of the fabrication of graphene oxide (GO)–bovines serum albumin (BSA) hybrid membrane for absorbing metal ions and reduction of metal ions.

2. Materials and Methods

2.1. Chemicals and Materials

Natural graphite flake (99.8% purity) was purchased from Sigma–Aldrich. Chloroauric acid ($HAuCl_4 \cdot 3H_2O$, ≥49.0% Au basis), disodium hydrogen phosphate (Na_2HPO_4), sodium dihydrogen phosphate (NaH_2PO_4), phosphoric acid (H_3PO_4), sulfuric acid (H_2SO_4), HCl, $NaBH_4$, potassium permanganate ($KMnO_4$), BSA, $ClCH_2COONa$, N-ethyl-N′-(3-dimethylaminopropyl) carbodiimide hydrochloride (EDC), and N-hydroxysuccinimide (NHS) were purchased from J&K company (Beijing, China). Ethanol was purchased from Beijing Chemicals Co., Ltd. (Beijing, China). All the chemicals used are at least of analytical grade and directly used without additional purification. The water purified through a Millipore system (18.2 MΩ·cm) was used throughout.

2.2. Synthesis of GO–BSA Nanohybrids

Graphene oxide was prepared from natural graphite flake by a modified Hummers method as described in our previous reports [22]. At first, in order to convert the ester group, hydroxyl group and epoxy group on the surface of GO to carboxyl group, GO was blended with $ClCH_2COONa$ and HCl. Briefly, 2 g NaOH and 2 g $ClCH_2COONa$ were added to 50 mL GO solution of 2 mg/mL concentration simultaneously. After sonication for 2 h, the product was neutralized by 1M HCl, followed by repeated centrifugation and rinsing to obtain GO–COOH products. The precipitation was again dispersed to 100 mL PBS (phosphate buffer saline) buffer (pH = 6.1). The second step is to make BSA react with GO–COOH, mainly by using $-NH_2$ of BSA and –COOH of GO–COOH. The specific operation was as follows, 50 mL GO–COOH (1 mg/mL) was dissolved in the PBS buffer solution. Then 20 mM EDC and 5 mM NHS were added and kept stirring for 1 h at room temperature. After that, 100 mg BSA in 10 mL PBS buffer solution (pH = 6.1) was added in the solution. Finally, the product was centrifuged and rinsed for three times, and then dialyzed for three days. The water was replaced three times a day to remove the inorganic ions. All the steps were conducted under room temperature to remove the unreacted BSA, EDC, NHS, and the ions in the buffer solution.

2.3. Fabrication of GO–BSA Membranes

The GO and GO–BSA membranes were fabricated by the well-introduced vacuum filtration method previously described [7,8]. In brief, GO and GO–BSA solutions were diluted to 1 mg/mL before filtrating. The as-prepared solutions were filtered through a 0.22 μm polyethersulfone (PES) membrane and then dried under 60 °C. The GO membrane and GO–BSA membrane were obtained with a load of about 5 mg/cm^2.

2.4. Removing Metallic Ions with Membranes

The absorbance of the solution before and after the membrane filtration was measured by using an ultraviolet spectrophotometer (PerkinElmer, America). 5 mM $HAuCl_4$, 0.1 M $CoCl_2$, $NiCl_2$, and $CuCl_2$ were configured, respectively. Metal ionic solutions were extracted with a 5-mL syringe containing the GO–BSA membranes. The filtrate was collected with a 2-mL centrifuge tube. The absorbance of the solution before and after filtration was measured several times. The GO–BSA membranes should maintain the same concentration and volume during the experiment to reduce experimental errors.

2.5. Characterization Techniques

A scanning electron microscope (SEM, JSM-6700F, JEOL) and a transmission electron microscope (TEM) were used to characterize the surface morphology changes before and after GO and BSA hybridization. The preparation procedures of the SEM samples were: dry GO powder, attach it to conductive adhesive, and then tape the GO–BSA hybrid membrane onto the conductive adhesive. The TEM images were acquired on a JEM-2100F field transmission electron microscopy (Hitachi

Limited, Japan) at an acceleration voltage of 200 kV. To prepare the TEM samples, the GO and GO–BSA were dispersed in deionized water by ultrasound for 1 h, and then a 5 μL solution was dripped on the copper grid with natural drying. The cpper grids for TEM characterization were purchased from the Plano GmbH (Wetzlar, Germany). Atom force microscopy (AFM) images were acquired on Bruker MultiMode 8 (Bruker, Beijing, China) in tapping mode. The type of AFM probe was a standard silicon tips (RTESP, Bruker, Beijing, China) with a spring constant of 40 N/m. The silicon wafer used for AFM was cleaned and dried with acetone and deionized water. The samples were diluted to a certain concentration and dripped on the silicon wafer with naturally drying. A Raman spectrometer (HORIBA Scientific, France) was used to characterize the sample using a laser source with a center wavelength of 780 nm. In the Raman spectroscopy measurement, the solid sample was placed directly on the quartz glass. The absorbance of the solution before and after the membrane filtration was measured using an ultraviolet spectrophotometer (PerkinElmer, America). X-ray diffraction (XRD) was taken on a Rigaku D/max-2500 VB+/PC equipment (BRUKER AXS GMBH, Germany) with a scanning 2θ angle of 5° to 90° at a voltage of 40 kV and a current of 40 MA.

3. Results and Discussion

3.1. Morphological Characterization of GO–BSA Membranes

To increase the efficiency of the conjugation between GO and BSA, we increased the content of the carboxyl group on the GO surface according to a previous work [35]. The oxygen-containing functional groups such as epoxy, ester, and hydroxyl on the surface of GO were effectively converted to –COOH by adding $ClCH_2COONa$ under alkaline conditions, as shown in Figure 2. It was reported that the $–NH_2$ contained in His, Arg, and Lys groups as well as the N-terminal of BSA can bind to –COOH of GO–COOH in the presence of EDC/NHS [41].

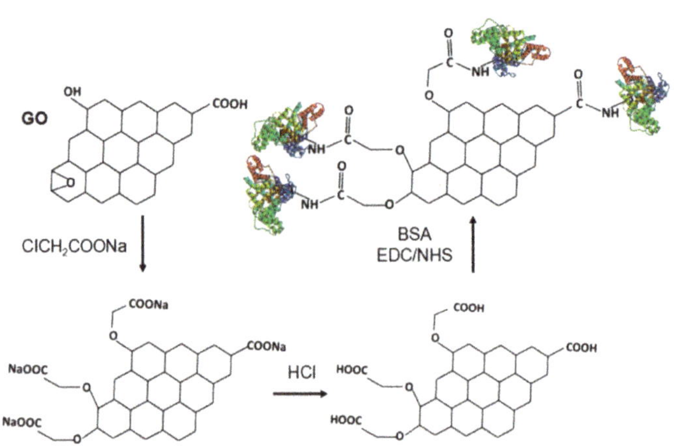

Figure 2. Schematic illustration of the functionalization of GO and the formation of GO–BSA nanohybrids by covalent interaction-mediated biomolecular self-assembly on GO surface.

The morphology and thickness of the fabricated GO–BSA membranes were first characterized by SEM, and the images are shown in Figure 3a,b. It can be found that the surface of the thin membranes presented some irregular holes, indicating that the fabricated membranes were not packed tightly. The cross-section of the GO–BSA membranes indicated the typical wrinkled lamellar structure. The thickness of the membranes was measured to be around 3 ± 1 μm. Due to the addition of protein molecules into the interlayer of GO, the loose structure inside the GO–BSA could be seen from the section diagram.

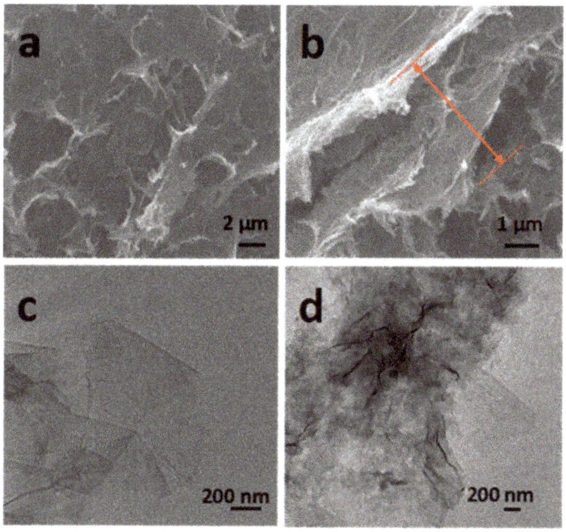

Figure 3. Morphological analysis of GO–BSA membranes: (**a**) SEM image of surface morphology of the GO–BSA film. (**b**) SEM image of the cross-section of the GO–BSA. (**c**) TEM image of surface morphology of the GO. (**d**) TEM image of surface morphology of the GO–BSA.

For comparison, the GO membrane was prepared with the same technique. The TEM image in Figure 3c showed that the GO, prepared by the modified Hummers method, displayed a thin lamellar structure with an area of hundreds of nanometers. When combined with BSA, the atoms on the surface caused a change of light and shade contrast. It can be clearly seen in Figure 3d that the surface of GO was transformed from the original smooth state into a rough surface structure. It indicates that a thin layer of protein was adsorbed onto the GO surface, which confirmed the successful functionalization of GO with BSA.

3.2. Structural and Property Characterizations of GO–BSA Membranes

The Raman spectrum was further used to characterize the change of the functional groups before and after the GO hybridization with BSA, as shown in Figure 4a. The Raman spectrums were measured by using a laser source with the center wavelength of 780 nm. Two peaks at 1315 cm^{-1} and 1585 cm^{-1} can be assigned to the D and G bands, respectively. The D band is related to the vibrations of sp^3 carbon atoms of disordered GO nanosheets, and the G band corresponded to the vibrations of sp^2 carbon atom domains of graphite. The GO–COOH Raman spectra exhibited that there was enhanced D peak at 1380 cm^{-1} (Figure 4a) and G-peak at 1576 cm^{-1} (Figure 4a), as well as no 2D peak at 2700 cm^{-1} (not shown) in the Raman spectrum, indicating that all GO nanosheets were functionalized as GO–COOH [42]. Furthermore, the intensity ratio between D and G bands (I_D/I_G) of GO was 0.95 and further increased to 1.01 for GO–COOH. This trend indicated the presence of disordered structures and defect sites on GO–COOH sheets, most likely resulting from the carboxylation of GO [43]. After the combination with BSA, the peak of the D at 1315 cm^{-1} and the G at 1585 cm^{-1} decreased compared with GO–COOH. The I_D/I_G increased from 1.01 to 1.11, indicating the defect density increases as a result of exfoliation and chemical modification [44,45]. Therefore, we suggested that BSA could not only be inserted into the GO nanosheets but can also interconnect GO monolayers together to form condensed GO–BSA hybrid membranes.

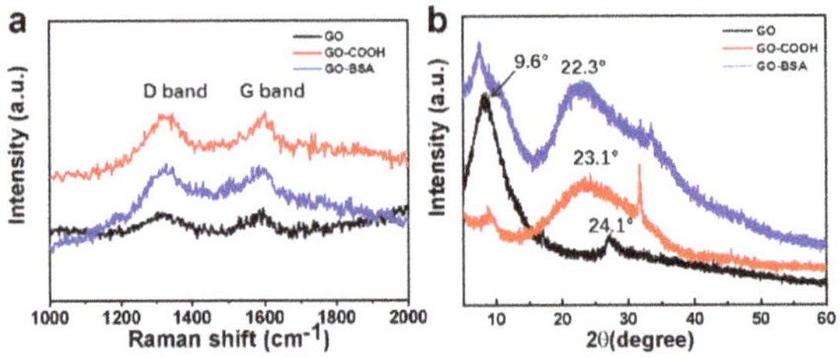

Figure 4. Structural and property characterizations of GO–BSA membranes: (**a**) Raman spectra of GO and GO–BSA hybrid membrane. (**b**) XRD patterns of the GO and GO–BSA hybrid membrane.

The atomic and molecular structures of GO, GO–COOH, and GO–BSA hybrid membranes were determined by XRD, shown in Figure 4b. The characteristic diffraction peak at 9.6° represents the (001) planes of GO [46]. A diffraction peak located at approximately 9.6° in XRD pattern of GO–COOH was also observed, corresponding to the crystal plane (001) of exfoliated GO [43]. After the formation of GO–BSA composite, the (001) planes still existed but seemed to be weaker compared with GO, illustrating the change of oxygen-containing groups. The broad peak of (002) planes located at 24.1° indicated the disordered stacking structure of the GO layers. The same peak located at 22.3° in the GO–BSA composites demonstrated that the GO–BSA composite had much more disordered stacking structures than GO and GO–COOH.

3.3. Separation of Metal Ions by GO–BSA Membranes

The BSA contained more than ten types of amino acids and was rich in lots of functional groups, such as imidazolyl and thiol groups (–SH), making it easy to adsorb heavy metal ions [40]. It had been reported that there was an interaction between BSA and gold ions involving biomineralization or biomimetic mineralization [9,37]. Since BSA contained 21 Tyr residues and possibly other residues with reduction functionality, it can reduce Au^{3+} through their phenolic groups, and their reducing power can be improved by adjusting the pH. For example, $AuCl_4^-$ can be reduced to Au(0) by biomolecules under alkaline conditions or in the presence of $NaBH_4$. Therefore, the prepared GO–BSA membranes are assumed to adsorb $AuCl_4^-$ in water and can reduce it to AuNPs.

As shown in Figure 5a, the UV profiles were performed based on different concentrations of $HAuCl_4$ solution. It can be seen that $HAuCl_4$ exhibited obvious absorption peaks at 310 nm, mainly due to the electron transition of Cl atoms to Au. The relationship between the absorbance and the concentration was fitted in Figure 5b. The linear correlation coefficient was about 0.99476, showing a good linear relationship. After the first and second rounds of the $HAuCl_4$ filtration (Figure 5c), there was almost no absorption intensity at 310 nm. This indicated that the GO–BSA membrane can effectively adsorb $AuCl_4^-$. But a weak absorption peak appears at 310 nm after the third cycle. Due to the partial detachment of $AuCl_4^-$ from the membrane and the increased concentration of $AuCl_4^-$, leading to the occurrence of characteristic peaks. According to the standard curve in Figure 5b, it can be calculated that the concentration of $AuCl_4^-$ after each filtration was 0.71 mM of the first filtration, 1.07 mM of the second filtration, and 1.43 mM of the third filtration, respectively. Particularly, for the first filtration, the efficiency was measured to up to 85 ± 1%. For the third filtration, part of the $AuCl_4^-$ fell off from the membrane, resulting in the increase of $AuCl_4^-$ concentration in the filtrate (light yellow in color), as shown in Figure 5d. Although GO had a certain adsorption capacity for $AuCl_4^-$, it can be seen from the Figure 5d that the GO–BSA had a better adsorption effect than it, which indicated that BSA had a better adsorption capacity for $AuCl_4^-$. In addition, although the stability of the membrane was relatively well under normal conditions, the performance of the GO–BSA membrane will be affected at high or low temperature due to the biological molecules it contained. Moreover, the GO–BSA membrane was also vulnerable to large impact.

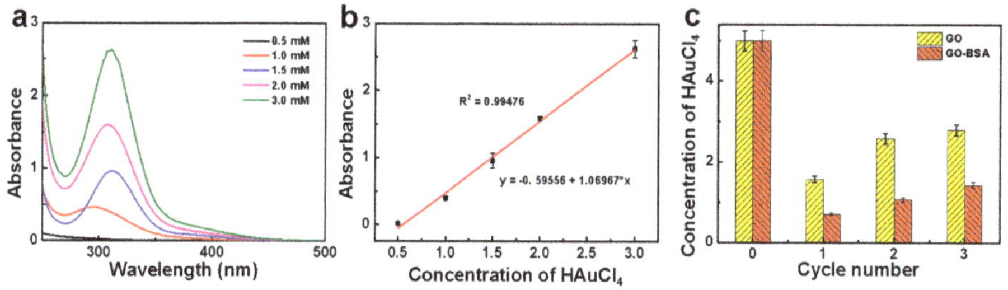

Figure 5. Removing $AuCl_4^-$ ions: (**a**) UV-Vis spectra of the $HAuCl_4$ with different concentrations, (**b**) the relationship between UV absorbance and $HAuCl_4$ concentration at 310 nm, and (**c**) calculation of the concentration of $HAuCl_4$ after the filtration according to the standard curve. Inset: the illustration of the color change before and after the filtration of the GO–BSA membrane.

The separation performances of several metal cations were determined via a similar method. The UV absorption of $CoCl_2$ appears at 511 nm (Figure 6a), which mainly resulted from the charge transfer between Co^{2+} and Cl^-. For the first filtration, the absorbance intensity was significantly decreased, demonstrating the $CoCl_2$ was adsorbed by the GO–BSA membrane. For the second filtration, Co^{2+} started to desorb from the membrane due to some weak binding between Co^{2+} and GO-BSA (Figure 6b). For the further filtration round, the absorbance was almost unchanged, indicating that the adsorption of Co^{2+} on the membrane achieved saturated. Based on the relationship between the absorbance and concentration of $CoCl_2$, a standard curve can be drawn (Figure 6c) with good linear fitting. The adsorption capacity of each filtration can be calculated according to the linear equation. Compared with the GO membrane, the adsorption capacity of GO–BSA at the first filtration was greater (Figure 6d), indicating that the functional groups on the protein had a certain effect on the binding of Co^{2+}.

In order to demonstrate the selective adsorption of the GO–BSA membrane towards Co^{2+}, we explored the adsorption capacity of the GO–BSA membranes to other metal ions, including Cu^{2+} and Fe^{2+}. Figure 6e,f present the adsorption capacities of the GO and GO–BSA membranes to Cu^{2+} and Fe^{2+}, respectively. This obtained results indicated that the GO–BSA membranes had no significant selective adsorption to Cu^{2+} and Fe^{2+}. In this process, BSA, as a metal binding protein, played the role of ion transporter. According to previous reports [35], Co^{2+} easily forms a stable combination with carboxylate and nitrogen, such as aspartate (Asp, D). While Cu^{2+} leaned toward nitrogen and sulfur ligands, such as histidine (His, H), cysteine (Cys, C), and methionine (Met, M) amino acid ligands. In addition, the similar adsorption amount of Cu^{2+} and Fe^{2+} may be related to ligand coordination geometry. The Co^{2+} ions showed a preference for octahedral while Cu^{2+} ions were easier to bind square-planar coordination geometry. The certain functional groups on GO–BSA, such as imidazole, can combine with Co^{2+} but not Cu^{2+} and Fe^{2+}, leading to selective adsorption of Co^{2+}. After several cycles, the adsorbed metal ions may fall off and enter into the filtrate, which led to the decrease of the adsorption capacity after the first cycle.

The metal ions adsorption performances of GO–BSA membranes were significantly enhanced attributing to the abundant functional groups on BSA molecules. Recent work by Zhang et al. [47] fabricated red-blood-cells like $BSA/Zn_3(PO_4)_2$ hybrid particles via the one-step method based on coordination between BSA and zinc ion. The adsorption efficiency of the hybrid particles increased with time from 86.33% (5 min) to 98.9% (30 min). The adsorption capacity was 6.85 mg/g at optimal conditions. They demonstrated that the hybrid particles displayed excellent adsorption properties on Cu^{2+} due to the high amount of BSA and Zn ions coordinated in the particles. This result proved the great potential of BSA-based nanomaterials applied in heavy metal ion treatment. A similar trend was presented in our system: GO–BSA membranes. Interestingly, the maximum adsorption capacity for Cu^{2+}, 0.3 mg/mg, was much larger than the hybrid particles in previous work. The higher adsorption capacity can be attributed to the combination of the large-surfaced GO and abundant functional groups of BSA molecules.

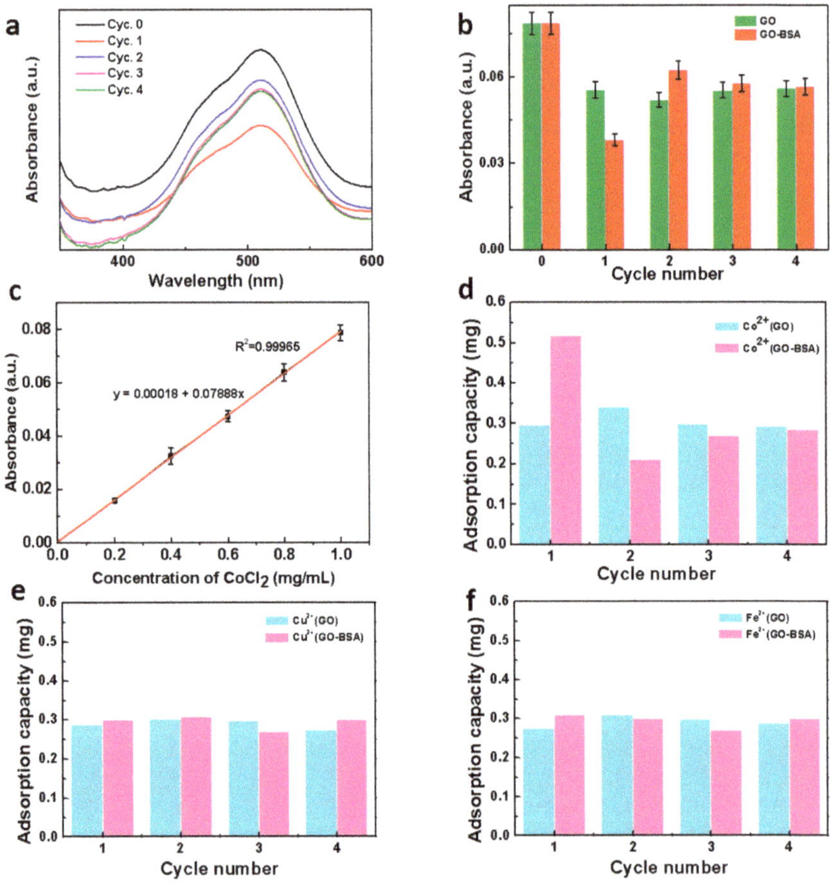

Figure 6. Removing other metallic ions: (**a**) the UV spectra of $CoCl_2$ before and after adsorbing by GO–BSA membrane; (**b**) the absorbance of $CoCl_2$ at 511 nm before and after adsorbing by GO–BSA membrane; (**c**) the standard curve of $CoCl_2$ at 511 nm; (**d**) the adsorption capacity of GO and GO–BSA filter membrane to $CoCl_2$; (**e**) the adsorption capacity of GO and GO–BSA filter membrane to $CuCl_2$. (**f**) The adsorption capacity of GO and GO–BSA membrane to $FeCl_2$.

3.4. Reduction of Au^{3+} on GO–BSA Membranes

The hybrid membrane was also assumed to reduce the adsorbed metallic ions to metal NPs. As a high-resolution imaging technique, AFM uses the relationship between the probe and the sample to detect the surface topography of the sample and obtain the size of AuNPs. Compared to pure GO with a thickness of around 1 nm (Figure 7a,e), the GO–BSA nanosheets exhibit a larger thickness within 20 nm. The significant increase of thickness can be attributed to the immobilization of BSA molecules on the GO–COOH nanosheets. Bovine serum albumin, as a serum albumin protein, displays globular conformation with a diameter of 14 nm under physiological conditions [48]. The increase of the nanosheet thickness after treatment with BSA indicates that lots of BSA molecules are immobilized on the surface and edges of GO–COOH (Figure 7b,f).

As was shown by AFM results in Figure 7b,c, the nanosheet thicknesses of the BSA immobilized GO–COOH and the GO–BSA–Au significantly increased. Meanwhile, the homogenous layer of BSA and BSA–Au can be observed on the surfaces of GO–COOH. This indicates that the saturation time of two hours is sufficient for GO–COOH fully-covered with BSA. Due to the limited COOH group on GO surfaces, the BSA immobilized on the GO–COOH surfaces contains not only covalently-bonded BSA molecules but also irreversibly adsorbed BSA molecules.

After treatment with $NaBH_4$, the thickness of GO–BSA–Au was measured to be around 30 nm. This indicates a large number of AuNPs grow on the surface of the GO–BSA nanosheets. The formation of AuNPs suggests that the adsorbed $AuCl_4^-$ was reduced to Au (Figure 7c). The selected profile of the corresponding AFM height image provides the dimension information of the GO sheets and AuNPs, as shown in Figure 7g. The size distribution of the AuNPs was relatively uniform (Figure 7h). Thus, the AuNPs that anchored on the surface of GO–BSA nanosheets have an average size of 17 $\pm$ 2 nm.

Liu et al. [49] adsorbed BSA onto GO/RGO and deposited Au nanoparticles on BSA using $HClO_4$. In comparison to their work, rather than assembling AuNPs directly onto the membrane, GO–BSA was a facile method without the involvement of latex (i.e., polystyrene sphere) assemblies. Besides, biological molecules in the GO–BSA membrane were not easy to desorb during the reduction of Au^{3+}. AuNPs can be stabilized by a combination of Au–S bonding with the protein (via the 35 Cys residues in BSA), and the steric protection due to the bulkiness of the protein [35]. The function of BSA in this process can also be confirmed by the work of Xie et al. [9], who reported that BSA can reduce $AuCl_4^-$ to Au (0) in an alkaline environment. The difference was that their reduced products were subnanometer-sized Au clusters, while most of the products we reduced were AuNPs. Besides, it can maintain a good adsorption capacity after repeated cycles. This indicated that the stability of the GO–BSA membrane was relatively well.

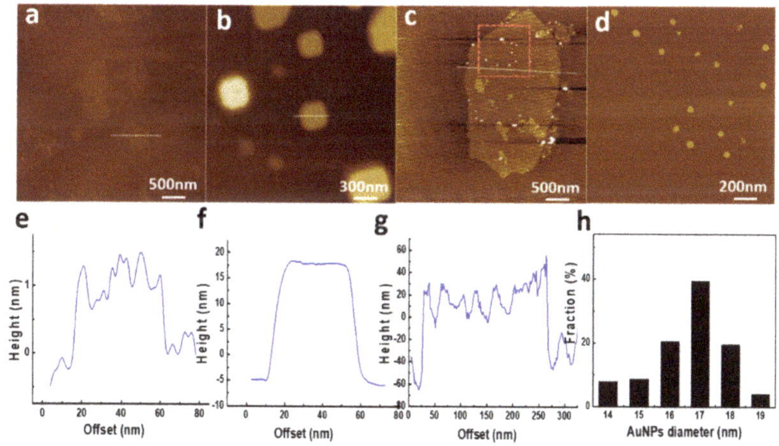

Figure 7. The AFM height images and corresponding section analysis of (**a**) GO, (**b**) BSA immobilized on GO, and (**c**) GO–BSA–Au, as well as their extracted line profiles of (**a**) GO, (**b**) BSA immobilized on GO, and (**c**) GO–BSA–Au, respectively. (**d**) AuNPs immobilized on the GO–BSA membrane. (**e**) Height image of extracted line profiles in (**a**) GO. (**f**) Height image of extracted line profiles in (**b**) BSA immobilized on GO. (**g**) Magnified height image of the red square area in (**c**), showing the AuNPs immobilized on the GO–BSA membrane. (**h**) The size distribution of AuNPs.

4. Conclusions

In summary, we reported that GO–BSA hybrid membranes can be used to remove metal ions from water. Due to the high metal-binding capacity of BSA and the large surface area of GO, the membrane can selectively absorb $AuCl_4^-$ from solutions. In addition, the reduction of $AuCl_4^-$ to AuNPs was successfully achieved after adding $NaBH_4$. This membrane was also demonstrated to selectively adsorb Co^{2+}, but not Cu^{2+} and Fe^{2+}. As a sewage treatment membrane, it was of great significance for removing heavy metal ions from water.

Author Contributions: X.Y. and S.S. designed and performed the experiments; L.Z. and X.Z. analyzed the data; Z.M. contributed experimental materials and analysis tools; G.W. optimized the data analysis; X.Y. and X.Z. wrote the paper under the supervision of Z.S. and G.W.

Funding: This research was funded by the National Natural Science Foundation of China (NSFC, grant no. 51573013, 51873016).

Acknowledgments: The authors gratefully acknowledge the financial support from the National Natural Science Foundation of China (NSFC, Grant no. 51573013, 51873016).

Conflicts of Interest: The authors declare no conflict of interest.

References

1. Shannon, M.A.; Bohn, P.W.; Elimelech, M.; Georgiadis, J.G.; Marinas, B.J.; Mayes, A.M. Science and technology for water purification in the coming decades. *Nature* **2008**, *452*, 301–310. [CrossRef] [PubMed]
2. Ivanets, A.I.; Kitikova, N.V.; Shashkova, I.L.; Oleksiienko, O.V.; Levchuk, I.; Sillanpää, M. Removal of Zn^{2+}, Fe^{2+}, Cu^{2+}, Pb^{2+}, Cd^{2+}, Ni^{2+} and Co^{2+} ions from aqueous solutions using modified phosphate dolomite. *J. Environ. Chem. Eng.* **2014**, *2*, 981–987. [CrossRef]
3. Lin, S.; Reddy, D.H.K.; Bediako, J.K.; Song, M.H.; Wei, W.; Kim, J.A.; Yun, Y.S. Effective adsorption of Pd(II), Pt(IV) and Au(III) by Zr(IV)-based metal-organic frameworks from strongly acidic solutions. *J. Mater. Chem. A* **2017**, *5*, 13557–13564. [CrossRef]
4. Zhang, W.S.; Yu, X.Q.; Li, Y.; Su, Z.Q.; Jandt, K.D.; Wei, G. Protein-mimetic peptide nanofibers: Motif design, self-assembly synthesis, and sequence-specific biomedical applications. *Prog. Polym. Sci.* **2018**, *80*, 94–124. [CrossRef]
5. Zhang, W.S.; Lin, D.M.; Wang, H.X.; Li, J.F.; Nienhaus, G.U.; Su, Z.Q.; Wei, G.; Shang, L. Supramolecular self-assembly bioinspired synthesis of luminescent gold nanocluster-embedded peptide nanofibers for temperature sensing and cellular imaging. *Bioconjugate Chem.* **2017**, *28*, 2224–2229. [CrossRef] [PubMed]
6. Wei, G.; Su, Z.Q.; Reynolds, N.P.; Arosio, P.; Hamley, I.W.; Gazit, E.; Mezzenga, R. Self-assembling peptide and protein amyloids: from structure to tailored function in nanotechnology. *Chem. Soc. Rev.* **2017**, *46*, 4661–4708. [CrossRef] [PubMed]
7. Bolisetty, S.; Reinhold, N.; Zeder, C.; Orozco, M.N.; Mezzenga, R. Efficient purification of arsenic-contaminated water using amyloid-carbon hybrid membranes. *Chem. Commun.* **2017**, *53*, 5714–5717. [CrossRef] [PubMed]
8. Bolisetty, S.; Mezzenga, R. Amyloid-carbon hybrid membranes for universal water purification. *Nat. Nanotechnol.* **2016**, *11*, 365–367. [CrossRef]
9. Xie, J.P.; Zheng, Y.G.; Ying, J.Y. Protein-directed synthesis of highly fluorescent gold nanoclusters. *J. Am. Chem. Soc.* **2009**, *131*, 888–889. [CrossRef]
10. Lu, F.; Zhang, S.H.; Gao, H.J.; Jia, H.; Zheng, L.Q. Protein-decorated reduced oxide graphene composite and its application to SERS. *ACS Appl. Mater. Interfaces* **2012**, *4*, 3278–3284. [CrossRef]
11. Stankovich, S.; Dikin, D.A.; Dommett, G.H.B.; Kohlhaas, K.M.; Zimney, E.J.; Stach, E.A.; Piner, R.D.; Nguyen, S.T.; Ruoff, R.S. Graphene-based composite materials. *Nature* **2006**, *442*, 282–286. [CrossRef] [PubMed]
12. Bae, S.; Kim, H.; Lee, Y.; Xu, X.F.; Park, J.S.; Zheng, Y.; Balakrishnan, J.; Lei, T.; Kim, H.R.; Song, Y.I.; et al. Roll-to-roll production of 30-inch graphene films for transparent electrodes. *Nat. Nanotechnol.* **2010**, *5*, 574–578. [CrossRef] [PubMed]
13. Wang, L.; Zhang, Y.J.; Wu, A.G.; Wei, G. Designed graphene-peptide nanocomposites for biosensor applications: A review. *Anal. Chim. Acta* **2017**, *985*, 24–40. [CrossRef] [PubMed]
14. Wang, L.; Wu, A.G.; Wei, G. Graphene-based aptasensors: from molecule-interface interactions to sensor design and biomedical diagnostics. *Analyst* **2018**, *143*, 1526–1543. [CrossRef] [PubMed]
15. Wang, Z.Q.; Ciacchi, L.C.; Wei, G. Recent advances in the synthesis of graphene-based nanomaterials for controlled drug delivery. *Appl. Sci.-Basel* **2017**, *7*, 1175. [CrossRef]

16. Zhang, H.C.; Gruner, G.; Zhao, Y.L. Recent advancements of graphene in biomedicine. *J. Mater. Chem. B* **2013**, *1*, 2542–2567. [CrossRef]
17. Wang, H.X.; Sun, D.M.; Zhao, N.N.; Yang, X.C.; Shi, Y.Z.; Li, J.F.; Su, Z.Q.; Wei, G. Thermo-sensitive graphene oxide-polymer nanoparticle hybrids: Synthesis, characterization, biocompatibility and drug delivery. *J. Mater. Chem. B* **2014**, *2*, 1362–1370. [CrossRef]
18. Li, X.L.; Zhi, L.J. Graphene hybridization for energy storage applications. *Chem. Soc. Rev.* **2018**, *47*, 3189–3216. [CrossRef]
19. Perreault, F.; de Faria, A.F.; Elimelech, M. Environmental applications of graphene-based nanomaterials. *Chem. Soc. Rev.* **2015**, *44*, 5861–5896. [CrossRef]
20. Li, A.H.; Liu, J.Q.; Feng, S.Y. Applications of graphene based materials in energy and environmental science. *Sci. Adv. Mater.* **2014**, *6*, 209–234. [CrossRef]
21. Zhao, X.N.; Zhang, P.P.; Chen, Y.T.; Su, Z.Q.; Wei, G. Recent advances in the fabrication and structure-specific applications of graphene-based inorganic hybrid membranes. *Nanoscale* **2015**, *7*, 5080–5093. [CrossRef] [PubMed]
22. Li, Y.; Zhang, P.P.; Ouyang, Z.F.; Zhang, M.F.; Lin, Z.J.; Li, J.F.; Su, Z.Q.; Wei, G. Nanoscale graphene doped with highly dispersed silver nanoparticles: Quick synthesis, facile fabrication of 3D membrane-modified electrode, and super performance for electrochemical sensing. *Adv. Funct. Mater.* **2016**, *26*, 2122–2134. [CrossRef]
23. Yu, X.Q.; Zhang, W.S.; Zhang, P.P.; Su, Z.Q. Fabrication technologies and sensing applications of graphene-based composite films: Advances and challenges. *Biosens. Bioelectron.* **2017**, *89*, 72–84. [CrossRef] [PubMed]
24. Zhang, H.; Li, Y.; Su, X.G. A small-molecule-linked DNA-graphene oxide-based fluorescence-sensing system for detection of biotin. *Anal. Biochem.* **2013**, *442*, 172–177. [CrossRef] [PubMed]
25. Li, Y.L.; Zhao, X.J.; Zhang, P.P.; Ning, J.; Li, J.F.; Su, Z.Q.; Wei, G. A facile fabrication of large-scale reduced graphene oxide-silver nanoparticle hybrid film as a highly active surface-enhanced Raman scattering substrate. *J. Mater. Chem. C* **2015**, *3*, 4126–4133. [CrossRef]
26. Yu, Y.; Liu, Y.; Zhen, S.J.; Huang, C.Z. A graphene oxide enhanced fluorescence anisotropy strategy for DNAzyme-based assay of metal ions. *Chem. Commun.* **2013**, *49*, 1942–1944. [CrossRef] [PubMed]
27. Zhu, Y.W.; Murali, S.; Cai, W.W.; Li, X.S.; Suk, J.W.; Potts, J.R.; Ruoff, R.S. Graphene and graphene oxide: Synthesis, properties, and applications. *Adv. Mater.* **2010**, *22*, 3906–3924. [CrossRef] [PubMed]
28. Yang, L.; Zhang, W.; Lu, Z.; Li, J.; Su, Z.; Gang, W. Sequence-designed peptide nanofibers bridged conjugation of graphene quantum dots with graphene oxide for high performance electrochemical hydrogen peroxide biosensor. *Adv. Mater. Interfaces* **2017**, *4*, 1600895.
29. Su, Z.; Shen, H.; Wang, H.; Wang, J.; Li, J.; Nienhaus, G.U.; Li, S.; Gang, W. Motif-designed peptide nanofibers decorated with graphene quantum dots for simultaneous targeting and imaging of tumor cells. *Adv. Funct. Mater.* **2015**, *25*, 5472–5478. [CrossRef]
30. Ali, I.; Alharbi, O.M.L.; Tkachev, A.; Galunin, E.; Burakov, A.; Grachev, V.A. Water treatment by new-generation graphene materials: Hope for bright future. *Environ. Sci. Pollut. R* **2018**, *25*, 7315–7329. [CrossRef] [PubMed]
31. Park, J.; Bazylewski, P.; Fanchini, G. Porous graphene-based membranes for water purification from metal ions at low differential pressures. *Nanoscale* **2016**, *8*, 9563–9571. [CrossRef] [PubMed]
32. Tabish, T.A.; Memon, F.A.; Gomez, D.E.; Horsell, D.W.; Zhang, S.W. A facile synthesis of porous graphene for efficient water and wastewater treatment. *Sci. Rep.* **2018**, *8*, 1817–1828. [CrossRef] [PubMed]
33. Suarez-Iglesias, O.; Collado, S.; Oulego, P.; Diaz, M. Graphene-family nanomaterials in wastewater treatment plants. *Chem. Eng. J.* **2017**, *313*, 121–135. [CrossRef]
34. Han, L.; Mao, D.; Huang, Y.C.; Zheng, L.J.; Yuan, Y.; Su, Y.; Sun, S.Y.; Fang, D. Fabrication of unique Tin(IV) Sulfide/Graphene Oxide for photocatalytically treating chromium(VI)-containing wastewater. *J. Clean. Prod.* **2017**, *168*, 519–525. [CrossRef]
35. Kim, S.; Nham, J.; Jeong, Y.S.; Lee, C.S.; Ha, S.H.; Park, H.B.; Lee, Y.J. Biomimetic selective ion transport through graphene oxide membranes functionalized with ion recognizing peptides. *Chem. Mater.* **2015**, *27*, 1255–1261. [CrossRef]
36. Zhang, M.F.; Li, Y.; Su, Z.Q.; Wei, G. Recent advances in the synthesis and applications of graphene-polymer nanocomposites. *Polym. Chem.* **2015**, *6*, 6107–6124. [CrossRef]

37. Yu, X.Q.; Wang, Z.P.; Su, Z.Q.; Wei, G. Design, fabrication, and biomedical applications of bioinspired peptide-inorganic nanomaterial hybrids. *J. Mater. Chem. B* **2017**, *5*, 1130–1142. [CrossRef]
38. Li, D.P.; Zhang, W.S.; Yu, X.Q.; Wang, Z.P.; Su, Z.Q.; Wei, G. When biomolecules meet graphene: From molecular level interactions to material design and applications. *Nanoscale* **2016**, *8*, 19491–19509. [CrossRef]
39. Liu, M.; Zhao, H.M.; Chen, S.; Yu, H.T.; Quan, X. Stimuli-responsive peroxidase mimicking at a smart graphene interface. *Chem. Commun.* **2012**, *48*, 7055–7057. [CrossRef]
40. Yu, X.Q.; Liu, W.; Deng, X.L.; Yan, S.Y.; Su, Z.Q. Gold nanocluster embedded bovine serum albumin nanofibers-graphene hybrid membranes for the efficient detection and separation of mercury ion. *Chem. Eng. J.* **2018**, *335*, 176–184. [CrossRef]
41. Chiu, N.F.; Fan, S.Y.; Yang, C.D.; Huang, T.Y. Carboxyl-functionalized graphene oxide composites as SPR biosensors with enhanced sensitivity for immunoaffinity detection. *Biosens. Bioelectron.* **2017**, *89*, 370–376. [CrossRef] [PubMed]
42. Lv, R.D.; Li, L.; Wang, Y.G.; Chen, Z.D.; Liu, S.C.; Wang, X.; Wang, J.; Li, Y.F. Carboxyl graphene oxide solution saturable absorber for femtosecond mode-locked erbium-doped fiber laser. *Chin. Phys. B* **2018**, *27*, 114214. [CrossRef]
43. Chen, J.; Zhang, X.; Cai, H.; Chen, Z.; Wang, T.; Jia, L.; Wang, J.; Wan, Q.; Pei, X. Osteogenic activity and antibacterial effect of zinc oxide/carboxylated graphene oxide nanocomposites: Preparation and in vitro evaluation. *Colloids Surf. B* **2016**, *147*, 397–407. [CrossRef] [PubMed]
44. Hung, Y.J.; Hofmann, M.; Cheng, Y.C.; Huang, C.W.; Chang, K.W.; Lee, J.Y. Characterization of graphene edge functionalization by grating enhanced Raman spectroscopy. *RSC Adv.* **2016**, *6*, 12398–12401. [CrossRef]
45. Zhao, X.J.; Li, Y.; Wang, J.H.; Ouyang, Z.F.; Li, J.F.; Wei, G.; Su, Z.Q. Interactive oxidation-reduction reaction for the in situ synthesis of graphene-phenol formaldehyde composites with enhanced properties. *ACS Appl. Mater. Interfaces* **2014**, *6*, 4254–4263. [CrossRef] [PubMed]
46. Stobinski, L.; Lesiak, B.; Malolepszy, A.; Mazurkiewicz, M.; Mierzwa, B.; Zemek, J.; Jiricek, P.; Bieloshapka, I. Graphene oxide and reduced graphene oxide studied by the XRD, TEM and electron spectroscopy methods. *J. Electron. Spectrosc. Relat. Phenom.* **2014**, *195*, 145–154. [CrossRef]
47. Zhang, B.; Li, P.; Zhang, H.; Li, X.; Tian, L.; Wang, H.; Chen, X.; Ali, N.; Ali, Z.; Zhang, Q. Red-blood-cell-like BSA/$Zn_3(PO_4)_2$ hybrid particles: preparation and application to adsorption of heavy metal ions. *Appl. Surf. Sci.* **2016**, *366*, 328–338. [CrossRef]
48. Wright, A.K.; Thompson, M.R. Hydrodynamic structure of BSA determined by transient electric birefringence. *Biophys. J.* **1975**, *15*, 137–141. [CrossRef]
49. Liu, J.B.; Fu, S.H.; Yuan, B.; Li, Y.L.; Deng, Z.X. Toward a universal "adhesive nanosheet" for the assembly of multiple nanoparticles based on a protein-induced reduction/decoration of graphene oxide. *J. Am. Chem. Soc.* **2010**, *132*, 7279–7281. [CrossRef]

MDPI
St. Alban-Anlage 66
4052 Basel
Switzerland
Tel. +41 61 683 77 34
Fax +41 61 302 89 18
www.mdpi.com

Nanomaterials Editorial Office
E-mail: nanomaterials@mdpi.com
www.mdpi.com/journal/nanomaterials

www.ingramcontent.com/pod-product-compliance
Lightning Source LLC
LaVergne TN
LVHW071957080526
838202LV00064B/6774